Student Solutions Manual

for

Thornton and Marion's

Classical Dynamics
of Particles and Systems

Fifth Edition

Stephen T. Thornton
University of Virginia

THOMSON

BROOKS/COLE

Australia • Canada • Mexico • Singapore • Spain • United Kingdom • United States

Contents

Preface

This manual contains the solutions to about 25% of the end-of-chapter problems from *Classical Dynamics of Particles and Systems*, 5th edition, by Stephen T. Thornton and Jerry B. Marion. The textbook also contains a large number of examples worked out in considerable detail in order to help the student learn not only the concepts, but good problem-solving techniques as well.

The student, whether he or she is studying science or engineering, should realize that it is important to be able to work problems in order to make sure the principles and concepts are understood. This is especially true in physics. Good problem-solving techniques cannot be just taught; they must be learned and assimilated through experience.

The student who reaches for this solution manual as soon as a homework problem is assigned will not learn much. The student should attempt to solve every assigned problem and only as a last result turn to this manual to obtain a *hint* for help. *It is very important for the student to attempt to work as many problems as possible.* This is not a manual to teach students how to work problems. The solutions here are condensed and shortened because of space limitations, and only the most important ideas and steps, leaving out some of the mathematics, are given. By putting in the missing steps where needed, the material presented here should allow the student to understand how to solve the problem.

Good students will attempt to work additional problems other than those assigned as homework, and the answers to all the even-numbered problems are given in the back of the textbook. If the student wants to solve additional problems in order to better learn the material, it might be a good idea to choose some of the problems with solutions given here. *After working out the problem, the student can then check this manual to see how they have done.* This solutions manual will allow a student working independently to monitor his or her progress.

What follows is a short list of problem-solving techniques that the student has undoubtedly seen earlier in the study of introductory physics. If these procedures are followed, there is a good chance that the problem can be solved.

Problem Solving Techniques:

1. Make a sketch or drawing of the problem to help visualize physically what is happening. Put parameters or variable names on the drawing.

2. If necessary, make another isolated sketch showing particularly important parts. Examples might be a free body diagram showing the forces on an object or a sketch showing only the generalized coordinates.

3. Write down all the known parameters of the problem, including the initial conditions and parameters known for any other time during the problem.

4. List what needs to be determined.

5. Think about the strategy needed to solve for the unknown. Write down equations that may be useful in solving for the unknown parameters. These may be kinematic equations, conservation laws, force laws, or Lagrange's equations, for example. These equations will undoubtedly have to be manipulated, sometimes several times, to obtain the needed answers.

6. Differentiation or integration will sometimes be needed, and often another parameter may have to be determined first in order to determine the particular parameter required. Multi-step problems are usually the most difficult, because they require the student to visualize an overall approach to the solution.

7. Sometimes it may be necessary (or easier) to solve the problem numerically. Take advantage of the several good computer software programs now available and learn to use one or two of them well.

8. It is normally best to not insert numerical values for parameters until the very end of the solution process. For example, it is best to use the symbol g for the acceleration of gravity until the end of the calculation, rather than to insert the value of 9.8 m/s^2 at the beginning.

The author (STT) would like to acknowledge the assistance of Tran ngoc Khanh (5th edition), Warren Griffith (4th edition), and Brian Giambattista (3rd edition), who checked the solutions of previous versions, went over user comments, and worked out solutions for new problems. Without their help, this manual would not be possible. Considerable thanks also go to Martin Arthur of Atelier 88 for typesetting all the problem solutions. The author would appreciate receiving reports of suggested improvements and suspected errors. Comments can be sent by email to stt@virginia.edu, the more detailed the better.

Stephen T. Thornton
Charlottesville, Virginia

Problems Solved in Student Solutions Manual

(the Student Solutions Manual is for sale to students)

Chapter	Problem Numbers
1	1, 2, 6, 9, 14, 17, 20, 26, 32, 35, 37
2	3, 7, 12, 17, 19, 23, 28, 29, 36, 41, 42, 49, 54
3	1, 6, 10, 14, 19, 24, 26, 29, 32, 37, 43
4	2, 4, 9, 13, 17, 20, 24
5	2, 5, 10, 16, 18
6	3, 7, 10, 14
7	3, 6, 10, 13, 16, 20, 23, 30, 32, 35, 41
8	3, 6, 12, 16, 23, 26, 31, 34, 39, 41, 46
9	3, 8, 15, 18, 20, 24, 28, 34, 35, 41, 45, 49, 55, 56, 62, 64
10	2, 8, 10, 14, 19
11	4, 6, 11, 14, 17, 22, 25, 31, 34
12	3, 6, 11, 14, 19, 23, 27
13	4, 7, 12, 13, 17
14	2, 8, 11, 14, 19, 24, 28, 31, 34, 36, 41

Solutions to all end-of-chapter problems are contained in the Instructor's Manual.

Matrices, Vectors, and Vector Calculus

1-1.

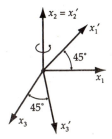

Axes x_1' and x_3' lie in the $x_1 x_3$ plane.

The transformation equations are:

$$x_1' = x_1 \cos 45° - x_3 \cos 45°$$

$$x_2' = x_2$$

$$x_3' = x_3 \cos 45° + x_1 \cos 45°$$

$$x_1' = \frac{1}{\sqrt{2}} x_1 - \frac{1}{\sqrt{2}} x_3$$

$$x_2' = x_2$$

$$x_3' = \frac{1}{\sqrt{2}} x_1 - \frac{1}{\sqrt{2}} x_3$$

So the transformation matrix is:

$$\begin{pmatrix} \dfrac{1}{\sqrt{2}} & 0 & -\dfrac{1}{\sqrt{2}} \\ 0 & 1 & 0 \\ \dfrac{1}{\sqrt{2}} & 0 & \dfrac{1}{\sqrt{2}} \end{pmatrix}$$

1-2.

a)

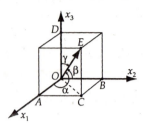

From this diagram, we have

$$\overline{OE} \cos \alpha = \overline{OA}$$

$$\overline{OE} \cos \beta = \overline{OB} \tag{1}$$

$$\overline{OE} \cos \gamma = \overline{OD}$$

Taking the square of each equation in (1) and adding, we find

$$\overline{OE}^2 \left[\cos^2 \alpha + \cos^2 \beta + \cos^2 \gamma \right] = \overline{OA}^2 + \overline{OB}^2 + \overline{OD}^2 \tag{2}$$

But

$$\overline{OA}^2 + \overline{OB}^2 = \overline{OC}^2 \tag{3}$$

and

$$\overline{OC}^2 + \overline{OD}^2 = \overline{OE}^2 \tag{4}$$

Therefore,

$$\overline{OA}^2 + \overline{OB}^2 + \overline{OD}^2 = \overline{OE}^2 \tag{5}$$

Thus,

$$\boxed{\cos^2 \alpha + \cos^2 \beta + \cos^2 \gamma = 1} \tag{6}$$

b)

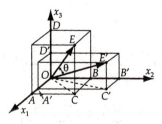

First, we have the following trigonometric relation:

$$\overline{OE}^2 + \overline{OE}'^2 - 2\overline{OE}\,\overline{OE}' \cos \theta = \overline{EE}'^2 \tag{7}$$

But,

$$\overline{EE}'^2 = \left[\overline{OB}' - \overline{OB}\right]^2 + \left[\overline{OA}' - \overline{OA}\right]^2 + \left[\overline{OD}' - \overline{OD}\right]^2$$

$$= \left[\overline{OE}' \cos \beta' - \overline{OE} \cos \beta\right]^2 + \left[\overline{OE}' \cos \alpha' - \overline{OE} \cos \alpha\right]^2$$

$$+ \left[\overline{OE}' \cos \gamma' - \overline{OE} \cos \gamma\right]^2 \tag{8}$$

or,

$$\overline{EE}'^2 = \overline{OE}'^2\left[\cos^2 \alpha' + \cos^2 \beta' + \cos^2 \gamma'\right] + \overline{OE}^2\left[\cos^2 \alpha + \cos^2 \beta + \cos^2 \gamma\right]$$

$$- 2\overline{OE}' \,\overline{OE}\left[\cos \alpha \cos \alpha' + \cos \beta \cos \beta' + \cos \gamma \cos \gamma'\right]$$

$$= OE'^2 + OE^2 - 2\overline{OE} \,\overline{OE}'\left[\cos \alpha \cos \alpha' + \cos \beta \cos \beta' + \cos \gamma \cos \gamma'\right] \tag{9}$$

Comparing (9) with (7), we find

$$\boxed{\cos \theta = \cos \alpha \cos \alpha' + \cos \beta \cos \beta' + \cos \gamma \cos \gamma'} \tag{10}$$

1-6. The lengths of line segments in the x_j and x_j' systems are

$$L = \sqrt{\sum_j x_j^2} \; ; \; L' = \sqrt{\sum_i x_i'^2} \tag{1}$$

If $L = L'$, then

$$\sum_j x_j^2 = \sum_i x_i'^2 \tag{2}$$

The transformation is

$$x_i' = \sum_j \lambda_{ij} x_j \tag{3}$$

Then,

$$\sum_j x_j^2 = \sum_i \left(\sum_k \lambda_{ik} x_k\right)\left(\sum_\ell \lambda_{i\ell} x_\ell\right)$$

$$= \sum_{k,\ell} x_k x_\ell \left(\sum_i \lambda_{ik} \lambda_{i\ell}\right) \tag{4}$$

But this can be true only if

$$\boxed{\sum_i \lambda_{ik} \lambda_{i\ell} = \delta_{k\ell}} \tag{5}$$

which is the desired result.

1-9. $A = i + 2j - k$ $B = -2i + 3j + k$

a)

$$\boxed{A - B = 3i - j - 2k}$$

$$|A - B| = \left[(3)^2 + (-1)^2 + (-2)^2 \right]^{1/2}$$

$$\boxed{|A - B| = \sqrt{14}}$$

b)

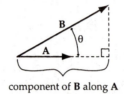

component of **B** along **A**

The length of the component of **B** along **A** is $B \cos \theta$.

$$A \cdot B = AB \cos \theta$$

$$B \cos \theta = \frac{A \cdot B}{A} = \frac{-2 + 6 - 1}{\sqrt{6}} = \frac{3}{\sqrt{6}} \text{ or } \frac{\sqrt{6}}{2}$$

The direction is, of course, along **A**. A unit vector in the **A** direction is

$$\frac{1}{\sqrt{6}}(i + 2j - k)$$

So the component of **B** along **A** is

$$\boxed{\frac{1}{2}(i + 2j - k)}$$

c) $$\cos \theta = \frac{A \cdot B}{AB} = \frac{3}{\sqrt{6}\sqrt{14}} = \frac{\sqrt{3}}{2\sqrt{7}} \, ; \; \theta = \cos^{-1} \frac{\sqrt{3}}{2\sqrt{7}}$$

$$\boxed{\theta \approx 71^\circ}$$

d) $$A \times B = \begin{vmatrix} i & j & k \\ 1 & 2 & -1 \\ -2 & 3 & 1 \end{vmatrix} = i \begin{vmatrix} 2 & -1 \\ 3 & 1 \end{vmatrix} - j \begin{vmatrix} 1 & -1 \\ -2 & 1 \end{vmatrix} + k \begin{vmatrix} 1 & 2 \\ -2 & 3 \end{vmatrix}$$

$$\boxed{A \times B = 5i + j + 7k}$$

e) $A - B = 3i - j - 2k$ $A + B = -i + 5j$

$$(A - B) \times (A + B) = \begin{vmatrix} i & j & k \\ 3 & -1 & -2 \\ -1 & 5 & 0 \end{vmatrix}$$

$$\boxed{(A - B) \times (A + B) = 10i + 2j + 14k}$$

1-14.

a)

$$\mathbf{AB} = \begin{bmatrix} 1 & 2 & -1 \\ 0 & 3 & 1 \\ 2 & 0 & 1 \end{bmatrix} \begin{bmatrix} 2 & 1 & 0 \\ 0 & -1 & 2 \\ 1 & 1 & 3 \end{bmatrix} = \begin{bmatrix} 1 & -2 & 1 \\ 1 & -2 & 9 \\ 5 & 3 & 3 \end{bmatrix}$$

Expand by the first row.

$$|\mathbf{AB}| = 1 \begin{vmatrix} -2 & 9 \\ 3 & 3 \end{vmatrix} + 2 \begin{vmatrix} 1 & 9 \\ 5 & 3 \end{vmatrix} + 1 \begin{vmatrix} 1 & -2 \\ 5 & 3 \end{vmatrix}$$

$$\boxed{|\mathbf{AB}| = -104}$$

b)

$$\mathbf{AC} = \begin{bmatrix} 1 & 2 & -1 \\ 0 & 3 & 1 \\ 2 & 0 & 1 \end{bmatrix} \begin{bmatrix} 2 & 1 \\ 4 & 3 \\ 1 & 0 \end{bmatrix} = \begin{bmatrix} 9 & 7 \\ 13 & 9 \\ 5 & 2 \end{bmatrix}$$

$$\boxed{\mathbf{AC} = \begin{bmatrix} 9 & 7 \\ 13 & 9 \\ 5 & 2 \end{bmatrix}}$$

c)

$$\mathbf{ABC} = \mathbf{A}(\mathbf{BC}) = \begin{bmatrix} 1 & 2 & -1 \\ 0 & 3 & 1 \\ 2 & 0 & 1 \end{bmatrix} \begin{bmatrix} 8 & 5 \\ -2 & -3 \\ 9 & 4 \end{bmatrix}$$

$$\boxed{\mathbf{ABC} = \begin{bmatrix} -5 & -5 \\ 3 & -5 \\ 25 & 14 \end{bmatrix}}$$

d)

$$\mathbf{AB} - \mathbf{B}^t \mathbf{A}^t = ?$$

$$\mathbf{AB} = \begin{bmatrix} 1 & -2 & 1 \\ 1 & -2 & 9 \\ 5 & 3 & 3 \end{bmatrix} \qquad \text{(from part } \mathbf{a})$$

$$\mathbf{B}^t \mathbf{A}^t = \begin{bmatrix} 2 & 0 & 1 \\ 1 & -1 & 1 \\ 0 & 2 & 3 \end{bmatrix} \begin{bmatrix} 1 & 0 & 2 \\ 2 & 3 & 0 \\ -1 & 1 & 1 \end{bmatrix} = \begin{bmatrix} 1 & 1 & 5 \\ -2 & -2 & 3 \\ 1 & 9 & 3 \end{bmatrix}$$

$$\boxed{\mathbf{AB} - \mathbf{B}^t \mathbf{A}^t = \begin{bmatrix} 0 & -3 & -4 \\ 3 & 0 & 6 \\ 4 & -6 & 0 \end{bmatrix}}$$

1-17.

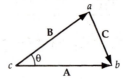

Consider the triangle a, b, c which is formed by the vectors **A**, **B**, **C**. Since

$$\mathbf{C} = \mathbf{A} - \mathbf{B}$$

$$|\mathbf{C}|^2 = (\mathbf{A} - \mathbf{B}) \cdot (\mathbf{A} - \mathbf{B}) \tag{1}$$

$$= A^2 - 2\mathbf{A} \cdot \mathbf{B} + B^2$$

or,

$$\boxed{|\mathbf{C}|^2 = A^2 + B^2 - 2AB \cos \theta} \tag{2}$$

which is the cosine law of plane trigonometry.

1-20.

a) Consider the following two cases:

When $i \neq j$ $\delta_{ij} = 0$ but $\varepsilon_{ijk} \neq 0$.

When $i = j$ $\delta_{ij} \neq 0$ but $\varepsilon_{ijk} = 0$.

Therefore,

$$\boxed{\sum_{ij} \varepsilon_{ijk} \, \delta_{ij} = 0} \tag{1}$$

b) We proceed in the following way:

When $j = k$, $\varepsilon_{ijk} = \varepsilon_{ijj} = 0$.

Terms such as $\varepsilon_{j11} \, \varepsilon_{\ell 11} = 0$. Then,

$$\sum_{jk} \varepsilon_{ijk} \, \varepsilon_{\ell jk} = \varepsilon_{i12} \, \varepsilon_{\ell 12} + \varepsilon_{i13} \, \varepsilon_{\ell 13} + \varepsilon_{i21} \, \varepsilon_{\ell 21} + \varepsilon_{i31} \, \varepsilon_{\ell 31} + \varepsilon_{i32} \, \varepsilon_{\ell 32} + \varepsilon_{i23} \, \varepsilon_{\ell 23}$$

Now, suppose $i = \ell = 1$, then,

$$\sum_{jk} = \varepsilon_{123} \, \varepsilon_{123} + \varepsilon_{132} \, \varepsilon_{132} = 1 + 1 = 2$$

for $i = \ell = 2$, $\displaystyle\sum_{jk} = \varepsilon_{213} \, \varepsilon_{213} + \varepsilon_{231} \, \varepsilon_{231} = 1 + 1 = 2$. For $i = \ell = 3$, $\displaystyle\sum_{jk} = \varepsilon_{312} \, \varepsilon_{312} + \varepsilon_{321} \, \varepsilon_{321} = 2$. But $i = 1$,

$\ell = 2$ gives $\displaystyle\sum_{jk} = 0$. Likewise for $i = 2$, $\ell = 1$; $i = 1$, $\ell = 3$; $i = 3$, $\ell = 1$; $i = 2$, $\ell = 3$; $i = 3$, $\ell = 2$.

Therefore,

$$\boxed{\sum_{j,k} \varepsilon_{ijk} \, \varepsilon_{\ell jk} = 2\delta_{i\ell}} \tag{2}$$

c) $\displaystyle\sum_{ijk} \varepsilon_{ijk} \, \varepsilon_{ijk} = \varepsilon_{123} \, \varepsilon_{123} + \varepsilon_{312} \, \varepsilon_{312} + \varepsilon_{321} \, \varepsilon_{321} + \varepsilon_{132} \, \varepsilon_{132} + \varepsilon_{213} \, \varepsilon_{213} + \varepsilon_{231} \, \varepsilon_{231}$

$$= 1\cdot 1 + 1\cdot 1 + (-1)\cdot(-1) + (-1)\cdot(-1) + (-1)\cdot(-1) + (1)\cdot(1)$$

or,

$$\boxed{\sum_{ijk} \varepsilon_{ijk} \, \varepsilon_{ijk} = 6} \tag{3}$$

1-26. When a particle moves along the curve

$$r = k(1 + \cos\theta) \tag{1}$$

we have

$$\left.\begin{array}{l} \dot{r} = -k\dot{\theta}\sin\theta \\[2mm] \ddot{r} = -k\left[\dot{\theta}^2 \cos\theta + \ddot{\theta}\sin\theta\right] \end{array}\right] \tag{2}$$

Now, the velocity vector in polar coordinates is [see Eq. (1.97)]

$$\mathbf{v} = \dot{r}\mathbf{e}_r + r\dot{\theta}\,\mathbf{e}_\theta \tag{3}$$

so that

$$v^2 = |\mathbf{v}|^2 = \dot{r}^2 + r^2\dot{\theta}^2$$

$$= k^2\dot{\theta}^2 \sin^2\theta + k^2\left(1 + 2\cos\theta + \cos^2\theta\right)\dot{\theta}^2$$

$$= k^2\dot{\theta}^2\left[2 + 2\cos\theta\right] \tag{4}$$

and v^2 is, by hypothesis, constant. Therefore,

$$\dot{\theta} = \sqrt{\frac{v^2}{2k^2\left(1 + \cos\theta\right)}} \tag{5}$$

Using (1), we find

$$\boxed{\dot{\theta} = \frac{v}{\sqrt{2kr}}} \tag{6}$$

Differentiating (5) and using the expression for $\dot{r}$, we obtain

$$\ddot{\theta} = \frac{v^2 \sin\theta}{4r^2} = \frac{v^2 \sin\theta}{4k^2\left(1 + \cos\theta\right)^2} \tag{7}$$

The acceleration vector is [see Eq. (1.98)]

$$\mathbf{a} = \left(\ddot{r} - r\dot{\theta}^2\right)\mathbf{e}_r + \left(r\ddot{\theta} + 2\dot{r}\dot{\theta}\right)\mathbf{e}_\theta \tag{8}$$

so that

$$\mathbf{a}\cdot\mathbf{e}_r = \ddot{r} - r\dot{\theta}^2$$

$$= -k\left(\dot{\theta}^2\cos\theta + \ddot{\theta}\sin\theta\right) - k\left(1 + \cos\theta\right)\dot{\theta}^2$$

$$= -k\left[\dot{\theta}^2\cos\theta + \frac{\ddot{\theta}^2\sin^2\theta}{2\left(1 + \cos\theta\right)} + \left(1 + \cos\theta\right)\dot{\theta}^2\right]$$

$$= -k\dot{\theta}^2\left[2\cos\theta + \frac{1 - \cos^2\theta}{2\left(1 + \cos\theta\right)} + 1\right]$$

$$= -\frac{3}{2}k\dot{\theta}^2\left(1 + \cos\theta\right) \tag{9}$$

or,

$$\boxed{\mathbf{a}\cdot\mathbf{e}_r = -\frac{3}{4}\frac{v^2}{k}} \tag{10}$$

In a similar way, we find

$$\mathbf{a}\cdot\mathbf{e}_\theta = -\frac{3}{4}\frac{v^2}{k}\frac{\sin\theta}{1 + \cos\theta} \tag{11}$$

From (10) and (11), we have

$$|\mathbf{a}| = \sqrt{\left(\mathbf{a}\cdot\mathbf{e}_r\right)^2 + \left(\mathbf{a}\cdot\mathbf{e}_\theta\right)^2} \tag{12}$$

or,

$$\boxed{|\mathbf{a}| = \frac{3}{4}\frac{v^2}{k}\sqrt{\frac{2}{1 + \cos\theta}}} \tag{13}$$

1-32.   Note that the integrand is a perfect differential:

$$2a\mathbf{r}\cdot\dot{\mathbf{r}} + 2b\dot{\mathbf{r}}\cdot\ddot{\mathbf{r}} = a\frac{d}{dt}(\mathbf{r}\cdot\mathbf{r}) + b\frac{d}{dt}(\dot{\mathbf{r}}\cdot\dot{\mathbf{r}}) \tag{1}$$

Clearly,

$$\boxed{\int\left(2a\mathbf{r}\cdot\dot{\mathbf{r}} + 2b\dot{\mathbf{r}}\cdot\ddot{\mathbf{r}}\right)dt = ar^2 + b\dot{r}^2 + \text{const.}} \tag{2}$$

1-35.

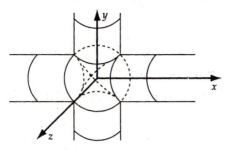

We compute the volume of the intersection of the two cylinders by dividing the intersection volume into two parts. Part of the common volume is that of one of the cylinders, for example, the one along the y axis, between $y = -a$ and $y = a$:

$$V_1 = 2\left(\pi a^2\right)a = 2\pi a^3 \tag{1}$$

The rest of the common volume is formed by 8 equal parts from the other cylinder (the one along the x-axis). One of these parts extends from $x = 0$ to $x = a$, $y = 0$ to $y = \sqrt{a^2 - x^2}$, $z = a$ to $z = \sqrt{a^2 - x^2}$. The complementary volume is then

$$V_2 = 8 \int_0^a dx \int_0^{\sqrt{a^2-x^2}} dy \int_a^{\sqrt{a^2-x^2}} dz$$

$$= 8 \int_0^a dx \sqrt{a^2 - x^2}\left[\sqrt{a^2 - x^2} - a\right]$$

$$= 8\left[a^2 x - \frac{x^3}{3} - \frac{a^3}{2}\sin^{-1}\frac{x}{a}\right]_0^a$$

$$= \frac{16}{3}a^3 - 2\pi a^3 \tag{2}$$

Then, from (1) and (2):

$$\boxed{V = V_1 + V_2 = \frac{16a^3}{3}} \tag{3}$$

1-37.

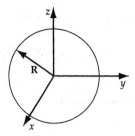

To do the integral directly, note that $\mathbf{A} = R^3 \mathbf{e}_r$, on the surface, and that $d\mathbf{a} = da\,\mathbf{e}_r$.

$$\int_S \mathbf{A} \cdot d\mathbf{a} = R^3 \int_S da = R^3 \times 4\pi R^2 = 4\pi R^5 \tag{1}$$

To use the divergence theorem, we need to calculate $\nabla \cdot \mathbf{A}$. This is best done in spherical coordinates, where $\mathbf{A} = r^3 \mathbf{e}_r$. Using Appendix F, we see that

$$\nabla \cdot \mathbf{A} = \frac{1}{r^2} \frac{\partial}{\partial r} \left(r^2 \mathbf{A}_r \right) = 5r^2 \tag{2}$$

Therefore,

$$\int_V \nabla \cdot \mathbf{A}\, dv = \int_0^\pi \sin\theta\, d\theta \int_0^{2\pi} d\phi \int_0^R r^2 \left(5r^2 \right) dr = 4\pi R^5 \tag{3}$$

Alternatively, one may simply set $dv = 4\pi r^2\, dr$ in this case.

Newtonian Mechanics— Single Particle

2-3.

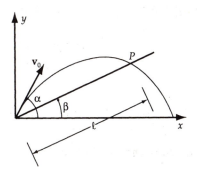

The equation of motion is

$$\mathbf{F} = m\,\mathbf{a} \tag{1}$$

The gravitational force is the only applied force; therefore,

$$\left. \begin{array}{l} F_x = m\ddot{x} = 0 \\[2mm] F_y = m\ddot{y} = -mg \end{array} \right] \tag{2}$$

Integrating these equations and using the initial conditions,

$$\left. \begin{array}{l} \dot{x}(t=0) = v_0 \cos\alpha \\[2mm] \dot{y}(t=0) = v_0 \sin\alpha \end{array} \right] \tag{3}$$

We find

$$\left. \begin{array}{l} \dot{x}(t) = v_0 \cos\alpha \\[2mm] \dot{y}(t) = v_0 \sin\alpha - gt \end{array} \right] \tag{4}$$

So the equations for x and y are

$$\left.\begin{aligned} x(t) &= v_0 t \cos\alpha \\ y(t) &= v_0 t \sin\alpha - \frac{1}{2} g t^2 \end{aligned}\right] \tag{5}$$

Suppose it takes a time t_0 to reach the point P. Then,

$$\left.\begin{aligned} \ell \cos\beta &= v_0 t_0 \cos\alpha \\ \ell \sin\beta &= v_0 t_0 \sin\alpha - \frac{1}{2} g t_0^2 \end{aligned}\right] \tag{6}$$

Eliminating ℓ between these equations,

$$\frac{1}{2} g t_0 \left(t_0 - \frac{2v_0 \sin\alpha}{g} + \frac{2v_0}{g} \cos\alpha \tan\beta \right) = 0 \tag{7}$$

from which

$$\boxed{t_0 = \frac{2v_0}{g}\left(\sin\alpha - \cos\alpha \tan\beta\right)} \tag{8}$$

2-7. Air resistance is always anti-parallel to the velocity. The vector expression is:

$$\mathbf{W} = \frac{1}{2} c_w \rho A v^2 \left[-\frac{\mathbf{v}}{v} \right] = -\frac{1}{2} c_w \rho A v \mathbf{v} \tag{1}$$

Including gravity and setting $\mathbf{F}_{net} = m\mathbf{a}$, we obtain the parametric equations

$$\ddot{x} = -b\dot{x}\sqrt{\dot{x}^2 + \dot{y}^2} \tag{2}$$

$$\ddot{y} = -b\dot{y}\sqrt{\dot{x}^2 + \dot{y}^2} - g \tag{3}$$

where $b = c_w \rho A/2m$. Solving with a computer using the given values and $\rho = 1.3\ \text{kg} \cdot \text{m}^{-3}$, we find that if the rancher drops the bale 210 m behind the cattle (the answer from the previous problem), then it takes $\simeq$ 4.44 s to land $\simeq$ 62.5 m behind the cattle. This means that the bale should be dropped at $\simeq$ 178 m behind the cattle to land 30 m behind. This solution is what is plotted in the figure. The time error she is allowed to make is the same as in the previous problem since it only depends on how fast the plane is moving.

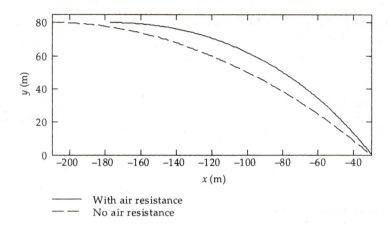

—— With air resistance
– – No air resistance

2-12. The equation of motion for the upward motion is

$$m \frac{d^2x}{dt^2} = -mkv^2 - mg \tag{1}$$

Using the relation

$$\frac{d^2x}{dt^2} = \frac{dv}{dt} = \frac{dv}{dx}\frac{dx}{dt} = v\frac{dv}{dx} \tag{2}$$

we can rewrite (1) as

$$\frac{v\,dv}{kv^2 + g} = -dx \tag{3}$$

Integrating (3), we find

$$\frac{1}{2k} \log\left(kv^2 + g\right) = -x + C \tag{4}$$

where the constant C can be computed by using the initial condition that $v = v_0$ when $x = 0$:

$$C = \frac{1}{2k} \log\left(kv_0^2 + g\right) \tag{5}$$

Therefore,

$$x = \frac{1}{2k} \log \frac{kv_0^2 + g}{kv^2 + g} \tag{6}$$

Now, the equation of downward motion is

$$m \frac{d^2x}{dt^2} = -mkv^2 + mg \tag{7}$$

This can be rewritten as

$$\frac{v\,dv}{-kv^2 + g} = dx \tag{8}$$

Integrating (8) and using the initial condition that $x = 0$ at $v = 0$ (w take the highest point as the origin for the downward motion), we find

$$x = \frac{1}{2k} \log \frac{g}{g - kv^2} \tag{9}$$

At the highest point the velocity of the particle must be zero. So we find the highest point by substituting $v = 0$ in (6):

$$x_h = \frac{1}{2k} \log \frac{kv_0^2 + g}{g} \tag{10}$$

Then, substituting (10) into (9),

$$\frac{1}{2k} \log \frac{kv_0^2 + g}{g} = \frac{1}{2k} \log \frac{g}{g - kv^2} \tag{11}$$

Solving for v,

$$v = \sqrt{\frac{\frac{g}{k} v_0^2}{v_0^2 + \frac{g}{k}}} \tag{12}$$

We can find the terminal velocity by putting $x \to \infty$ in (9). This gives

$$v_t = \sqrt{\frac{g}{k}} \tag{13}$$

Therefore,

$$\boxed{v = \frac{v_0 v_t}{\sqrt{v_0^2 + v_t^2}}} \tag{14}$$

2-17.

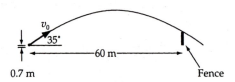

The setup for this problem is as follows:

$$x = v_0 t \cos \theta \tag{1}$$

$$y = y_0 + v_0 t \sin \theta - \frac{1}{2} g t^2 \tag{2}$$

where $\theta = 35°$ and $y_0 = 0.7 \text{ m}$. The ball crosses the fence at a time $\tau = R/(v_0 \cos \theta)$, where $R = 60$ m. It must be at least $h = 2$ m high, so we also need $h - y_0 = v_0 \tau \sin \theta - g\tau^2/2$. Solving for v_0, we obtain

$$v_0^2 = \frac{gR^2}{2\cos\theta\left[R\sin\theta - (h - y_0)\cos\theta\right]} \tag{3}$$

which gives $v_0 \simeq 25.4\ \mathrm{m\cdot s^{-1}}$.

2-19. The projectile's motion is described by

$$\left.\begin{array}{l} x = (v_0\cos\alpha)t \\[2mm] y = (v_0\sin\alpha)t - \dfrac{1}{2}gt^2 \end{array}\right] \tag{1}$$

where v_0 is the initial velocity. The distance from the point of projection is

$$r = \sqrt{x^2 + y^2} \tag{2}$$

Since r must always increase with time, we must have $\dot{r} > 0$:

$$\dot{r} = \frac{x\dot{x} + y\dot{y}}{r} > 0 \tag{3}$$

Using (1), we have

$$x\dot{x} + y\dot{y} = \frac{1}{2}g^2t^3 - \frac{3}{2}g(v_0\sin\alpha)t^2 + v_0^2 t \tag{4}$$

Let us now find the value of t which yields $x\dot{x} + y\dot{y} = 0$ (i.e., $\dot{r} = 0$):

$$t = \frac{3}{2}\frac{v_0\sin\alpha}{g} \pm \frac{v_0}{2g}\sqrt{9\sin^2\alpha - 8} \tag{5}$$

For small values of α, the second term in (5) is imaginary. That is, $r = 0$ is never attained and the value of t resulting from the condition $\dot{r} = 0$ is unphysical.

Only for values of α greater than the value for which the radicand is zero does t become a physical time at which $\dot{r}$ does in fact vanish. Therefore, the maximum value of α that insures $\dot{r} > 0$ for all values of t is obtained from

$$9\sin^2\alpha_{max} - 8 = 0 \tag{6}$$

or,

$$\sin\alpha_{max} = \frac{2\sqrt{2}}{3} \tag{7}$$

so that

$$\boxed{\alpha_{max} \cong 70.5°} \tag{8}$$

2-23.
$$F(t) = ma(t) = kte^{-at} \tag{1}$$

with the initial conditions $x(t) = v(t) = 0$. We integrate to get the velocity. Showing this explicitly,

$$\int_{v(0)}^{v(t)} a(t)\, dt = \frac{k}{m} \int_0^t te^{-\alpha t}\, dt \tag{2}$$

Integrating this by parts and using our initial conditions, we obtain

$$v(t) = \frac{k}{m}\left[\frac{1}{\alpha^2} - \frac{1}{\alpha}\left(t + \frac{1}{\alpha}\right)e^{-\alpha t}\right] \tag{3}$$

By similarly integrating $v(t)$, and using the integral (2) we can obtain $x(t)$.

$$x(t) = \frac{k}{m}\left[-\frac{2}{\alpha^3} + \frac{1}{\alpha^2} + \frac{1}{\alpha^2}\left(t + \frac{2}{\alpha}\right)e^{-\alpha t}\right] \tag{4}$$

To make our graphs, substitute the given values of $m = 1$ kg, $k = 1$ N·s^{-1}, and $\alpha = 0.5$ s^{-1}.

$$x(t) = te^{-t/2} \tag{5}$$

$$v(t) = 4 - 2(t + 2)e^{-t/2} \tag{6}$$

$$\alpha(t) = -16 + 4t + 4(t + 4)e^{-t/2} \tag{7}$$

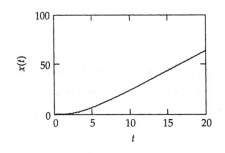

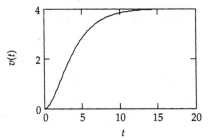

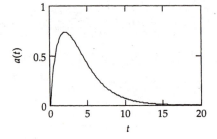

2-28.

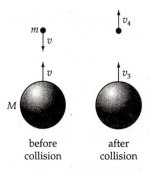

before
collision

after
collision

The problem, as stated, is completely one-dimensional. We may therefore use the elementary result obtained from the use of our conservation theorems: energy (since the collision is elastic) and momentum. We can factor the momentum conservation equation

$$m_1 v_1 + m_2 v_2 = m_1 v_3 + m_2 v_4 \tag{1}$$

out of the energy conservation equation

$$\frac{1}{2} m_1 v_1^2 + \frac{1}{2} m_2 v_2^2 = \frac{1}{2} m_1 v_3^2 + \frac{1}{2} m_2 v_4^2 \tag{2}$$

and get

$$v_1 + v_3 = v_2 + v_4 \tag{3}$$

This is the "conservation" of relative velocities that motivates the definition of the coefficient of restitution. In this problem, we initially have the superball of mass M coming up from the ground with velocity $v = \sqrt{2gh}$, while the marble of mass m is falling at the same velocity. Conservation of momentum gives

$$Mv + m(-v) = Mv_3 + mv_4 \tag{4}$$

and our result for elastic collisions in one dimension gives

$$v + v_3 = (-v) + v_4 \tag{5}$$

solving for v_3 and v_4 and setting them equal to $\sqrt{2gh_{item}}$, we obtain

$$h_{marble} = \left[\frac{3-\alpha}{1+\alpha} \right]^2 h \tag{6}$$

$$h_{superball} = \left[\frac{1-3\alpha}{1+\alpha} \right]^2 h \tag{7}$$

where $\alpha \equiv m/M$. Note that if $\alpha < 1/3$, the superball will bounce on the floor a second time after the collision.

2-29.

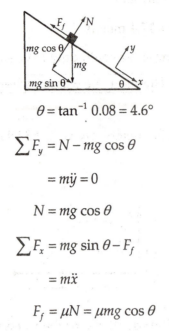

$$\theta = \tan^{-1} 0.08 = 4.6°$$

$$\sum F_y = N - mg \cos \theta$$

$$= m\ddot{y} = 0$$

$$N = mg \cos \theta$$

$$\sum F_x = mg \sin \theta - F_f$$

$$= m\ddot{x}$$

$$F_f = \mu N = \mu mg \cos \theta$$

so

$$m\ddot{x} = mg \sin \theta - \mu mg \cos \theta$$

$$\ddot{x} = g(\sin \theta - \mu \cos \theta)$$

Integrate with respect to time

$$\dot{x} = gt(\sin \theta - \mu \cos \theta) + \dot{x}_0 \qquad (1)$$

Integrate again:

$$x = x_0 + \dot{x}_0 t + \frac{1}{2} gt^2 (\sin \theta - \mu \cos \theta) \qquad (2)$$

Now we calculate the time required for the driver to stop for a given $\dot{x}_0$ (initial speed) by solving Eq. (1) for t with $\dot{x} = 0$.

$$t' = -\frac{\dot{x}_0}{g}(\sin \theta - \mu \cos \theta)^{-1}$$

Substituting this time into Eq. (2) gives us the distance traveled before coming to a stop.

$$(x' - x_0) = \dot{x}_0 t' + \frac{1}{2} gt'^2 (\sin \theta - \mu \cos \theta)$$

$$\Delta x = -\frac{\dot{x}_0^2}{g}(\sin \theta - \mu \cos \theta)^{-1} + \frac{1}{2} g \frac{\dot{x}_0^2}{g^2}(\sin \theta - \mu \cos \theta)^{-1}$$

$$\Delta x = \frac{\dot{x}_0^2}{2g}(\mu \cos \theta - \sin \theta)^{-1}$$

We have $\theta = 4.6°$, $\mu = 0.45$, $g = 9.8 \text{ m/sec}^2$.

For $\dot{x}_0 = 25 \text{ mph} = 11.2 \text{ m/sec}$, $\Delta x = 17.4 \text{ meters}$.

If the driver had been going at 25 mph, he could only have skidded 17.4 meters.

$\boxed{\text{Therefore, he was speeding}}$

How fast was he going?

$$\Delta x \geq 30 \text{ meters} \quad \text{gives} \quad \dot{x}_0 \geq 32.9 \text{ mph}.$$

2-36.

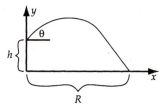

Put the origin at the initial point. The equations for the x and y motion are then

$$x = v_0 (\cos \theta) t$$

$$y = v_0 (\sin \theta) t - \frac{1}{2} g t^2$$

Call τ the time when the projectile lands on the valley floor. The y equation then gives

$$-h = v_0 (\sin \theta) \tau - \frac{1}{2} g \tau^2$$

Using the quadratic formula, we may find τ

$$\tau = \frac{v_0 \sin \theta}{g} + \frac{\sqrt{v_0^2 \sin^2 \theta + 2gh}}{g}$$

(We take the positive since $\tau > 0$.) Substituting τ into the x equation gives the range R as a function of θ.

$$R = \frac{v_0^2}{g} \cos \theta \left(\sin \theta + \sqrt{\sin^2 \theta + x^2} \right) \tag{1}$$

where we have defined $x^2 \equiv 2gh/v_0^2$. To maximize R for a given h and v_0, we set $dR/d\theta = 0$. The equation we obtain is

$$\cos^2 \theta - \sin^2 \theta - \sin \theta \sqrt{\sin^2 \theta + x^2} + \frac{\sin \theta \cos^2 \theta}{\sqrt{\sin^2 \theta + x^2}} = 0 \tag{2}$$

Although it can give $x = x(\theta)$, the above equation cannot be solved to give $\theta = \theta(x)$ in terms of the elementary functions. The optimum θ for a given x is plotted in the figure, along with its respective range in units of v_0^2/g. Note that $x = 0$, which among other things corresponds to $h = 0$, gives the familiar result $\theta = 45°$ and $R = v_0^2/g$.

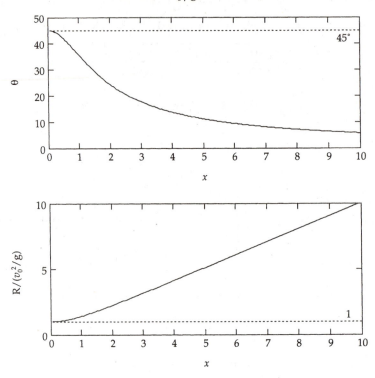

2-41.

a) As measured on the train:

$$T_i = 0; \; T_f = \frac{1}{2}mv^2$$

$$\boxed{\Delta T = \frac{1}{2}mv^2}$$

b) As measured on the ground:

$$T_i = \frac{1}{2}mu^2; \; T_f = \frac{1}{2}m(v+u)^2$$

$$\boxed{\Delta T = \frac{1}{2}mv^2 + mvu}$$

c) The woman does an amount of work equal to the kinetic energy gain of the ball as measured in her frame.

$$\boxed{W = \frac{1}{2}mv^2}$$

d) The train does work in order to keep moving at a constant speed u. (If the train did no work, its speed after the woman threw the ball would be slightly less than u, and the speed of the ball relative to the ground would not be $u + v$.) The term mvu is the work that must be supplied by the train.

$$\boxed{W = mvu}$$

2-42.

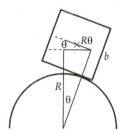

From the figure, we have $h(\theta) = (R + b/2)\cos\theta + R\theta\sin\theta$, and the potential is $U(\theta) = mgh(\theta)$. Now compute:

$$\frac{dU}{d\theta} = mg\left[-\frac{b}{2}\sin\theta + R\theta\cos\theta\right] \tag{1}$$

$$\frac{d^2U}{d\theta^2} = mg\left[\left(R - \frac{b}{2}\right)\cos\theta - R\theta\sin\theta\right] \tag{2}$$

The equilibrium point (where $dU/d\theta = 0$) that we wish to look at is clearly $\theta = 0$. At that point, we have $d^2U/d\theta^2 = mg(R - b/2)$, which is stable for $R > b/2$ and unstable for $R < b/2$. We can use the results of Problem 2-46 to obtain stability for the case $R = b/2$, where we will find that the first non-trivial result is in fourth order and is negative. We therefore have an equilibrium at $\theta = 0$ which is stable for $R > b/2$ and unstable for $R \le b/2$.

2-49. The distances from stars to the center of mass of the system are respectively

$$r_1 = \frac{dm_2}{m_1 + m_2} \quad \text{and} \quad r_2 = \frac{dm_1}{m_1 + m_2}$$

At equilibrium, like in previous problem, we have

$$\frac{Gm_1m_2}{d^2} = \frac{m_1v_1^2}{r_1} \Rightarrow v_1 = \sqrt{\frac{Gm_2^2}{d(m_1 + m_2)}} \Rightarrow \tau = \frac{2\pi r_1}{v_1} = \frac{2\pi d^{3/2}}{\sqrt{G(m_1 + m_2)}}$$

The result will be the same if we consider the equilibrium of forces acting on 2nd star.

2-54.

a) Terminal velocity means final steady velocity (here we assume that the potato reaches this velocity before the impact with the Earth) when the total force acting on the potato is zero.

$$mg = kmv \quad \text{and consequently} \quad v = g/k = 1000 \text{ m/s}.$$

b)

$$\frac{dv}{dt} = \frac{F}{m} = -(g + kv) \Rightarrow \frac{dx}{v} = dt = -\frac{dv}{g + kv} \Rightarrow \int_0^x dx = -\int_{v_0}^0 \frac{v\,dv}{g + kv} \Rightarrow$$

$$x_{max} = \frac{v_0}{k} + \frac{g}{k^2} \ln \frac{g}{g + kv_0} = 679.7 \text{ m} \quad \text{where } v_0 \text{ is the initial velocity of the potato.}$$

CHAPTER 3

Oscillations

3-1.

a)
$$\nu_0 = \frac{1}{2\pi}\sqrt{\frac{k}{m}} = \frac{1}{2\pi}\sqrt{\frac{10^4 \text{ dyne/cm}}{10^2 \text{ gram}}} = \frac{10}{2\pi}\sqrt{\frac{\dfrac{\text{gram} \cdot \text{cm}}{\text{sec}^2 \cdot \text{cm}}}{\text{gram}}} = \frac{10}{2\pi} \text{ sec}^{-1}$$

or,

$$\boxed{\nu_0 \cong 1.6 \text{ Hz}} \tag{1}$$

$$\tau_0 = \frac{1}{\nu_0} = \frac{2\pi}{10} \text{ sec}$$

or,

$$\boxed{\tau_0 \cong 0.63 \text{ sec}} \tag{2}$$

b)
$$E = \frac{1}{2}kA^2 = \frac{1}{2} \times 10^4 \times 3^2 \text{ dyne-cm}$$

so that

$$\boxed{E = 4.5 \times 10^4 \text{ erg}} \tag{3}$$

c)   The maximum velocity is attained when the total energy of the oscillator is equal to the kinetic energy. Therefore,

$$\frac{1}{2}mv_{max}^2 = 4.5 \times 10^4 \text{ erg}$$

$$v_{max} = \sqrt{\frac{2 \times 4.5 \times 10^4}{100}}$$

or,

$$\boxed{v_{\max} = 30 \text{ cm/sec}} \tag{4}$$

3-6.

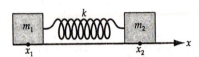

Suppose the coordinates of m_1 and m_2 are x_1 and x_2 and the length of the spring at equilibrium is ℓ. Then the equations of motion for m_1 and m_2 are

$$m_1 \ddot{x}_1 = -k(x_1 - x_2 + \ell) \tag{1}$$

$$m_2 \ddot{x}_2 = -k(x_2 - x_1 + \ell) \tag{2}$$

From (2), we have

$$x_1 = \frac{1}{k}(m_2 \ddot{x}_2 + kx_2 - k\ell) \tag{3}$$

Substituting this expression into (1), we find

$$\frac{d^2}{dt^2}\left[m_1 m_2 \ddot{x}_2 + (m_1 + m_2)kx_2\right] = 0 \tag{4}$$

from which

$$\ddot{x}_2 = -\frac{m_1 + m_2}{m_1 m_2}kx_2 \tag{5}$$

Therefore, x_2 oscillates with the frequency

$$\boxed{\omega = \sqrt{\frac{m_1 + m_2}{m_1 m_2}k}} \tag{6}$$

We obtain the same result for x_1. If we notice that the reduced mass of the system is defined as

$$\frac{1}{\mu} = \frac{1}{m_1} + \frac{1}{m_2} \tag{7}$$

we can rewrite (6) as

$$\omega = \sqrt{\frac{k}{\mu}} \tag{8}$$

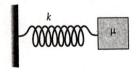

This means the system oscillates in the same way as a system consisting of a single mass μ. Inserting the given values, we obtain $\mu \approx 66.7$ g and $\omega \approx 2.74$ rad $\cdot$ s^{-1}.

3-10. The amplitude of a damped oscillator is expressed by

$$x(t) = Ae^{-\beta t} \cos(\omega_1 t + \delta) \tag{1}$$

Since the amplitude decreases to $1/e$ after n periods, we have

$$\beta n T = \beta n \frac{2\pi}{\omega_1} = 1 \tag{2}$$

Substituting this relation into the equation connecting ω_1 and ω_0 (the frequency of undamped oscillations), $\omega_1^2 = \omega_0^2 - \beta^2$, we have

$$\omega_0^2 = \omega_1^2 + \left[\frac{\omega_1}{2\pi n}\right]^2 = \omega_1^2 \left[1 + \frac{1}{4\pi^2 n^2}\right] \tag{3}$$

Therefore,

$$\frac{\omega_1}{\omega_0} = \left[1 + \frac{1}{4\pi^2 n^2}\right]^{-1/2} \tag{4}$$

so that

$$\boxed{\frac{\omega_1}{\omega_2} \cong 1 - \frac{1}{8\pi^2 n^2}}$$

3-14. For the case of overdamped oscillations, $x(t)$ and $\dot{x}(t)$ are expressed by

$$x(t) = e^{-\beta t}\left[A_1 e^{\omega_2 t} + A_2 e^{-\omega_2 t}\right] \tag{1}$$

$$\dot{x}(t) e^{-\beta t}\left[-\beta\left(A_1 e^{\omega_2 t} + + A_2 e^{-\omega_2 t}\right) + \left(A_1 \omega_2 e^{\omega_2 t} - A_2 \omega_2 e^{-\omega_2 t}\right)\right] \tag{2}$$

where $\omega_2 = \sqrt{\beta^2 - \omega_0^2}$. Hyperbolic functions are defined as

$$\cosh y = \frac{e^y + e^{-y}}{2}, \quad \sinh y = \frac{e^y - e^{-y}}{2} \tag{3}$$

or,

$$\left.\begin{array}{l} e^y = \cosh y + \sinh y \\[2mm] e^{-y} = \cosh y - \sinh y \end{array}\right] \tag{4}$$

Using (4) to rewrite (1) and (2), we have

$$\boxed{x(t) = (\cosh \beta t - \sinh \beta t)\left[(A_1 + A_2)\cosh \omega_2 t + (A_1 - A_2)\sinh \omega_2 t\right]} \tag{5}$$

and

$$\dot{x}(t) = (\cosh \beta t - \sinh \beta t)\Big[\big(A_1\omega_2 - A_1\beta\big)\big(\cosh \omega_2 t + \sinh \omega_2 t\big)$$
$$- \big(A_2\beta + A_2\omega_2\big)\big(\cosh \omega_2 t - \sinh \omega_2 t\big)\Big] \tag{6}$$

3-19. The amplitude of a damped oscillator is [Eq. (3.59)]

$$D = \frac{A}{\sqrt{\left(\omega_0^2 - \omega^2\right)^2 + 4\omega^2\beta^2}} \tag{1}$$

At the resonance frequency, $\omega = \omega_R = \sqrt{\omega_0^2 - \beta^2}$, D becomes

$$D_R = \frac{A}{2\beta\sqrt{\omega_0^2 - \beta^2}} \tag{2}$$

Let us find the frequency, $\omega = \omega'$, at which the amplitude is $\dfrac{1}{\sqrt{2}}D_R$:

$$\frac{1}{\sqrt{2}}D_R = \frac{1}{\sqrt{2}}\frac{A}{2\beta\sqrt{\omega_0^2 - \beta^2}} = \frac{A}{\sqrt{\left(\omega_0^2 - \omega'^2\right)^2 + 4\omega'^2\beta^2}} \tag{3}$$

Solving this equation for ω', we find

$$\omega'^2 = \omega_0^2 - 2\beta^2 \pm 2\beta\omega_0\left[1 - \frac{\beta^2}{\omega_0^2}\right]^{1/2} \tag{4}$$

For a lightly damped oscillator, β is small and the terms in β^2 can be neglected. Therefore,

$$\omega'^2 \cong \omega_0^2 \pm 2\beta\omega_0 \tag{5}$$

or,

$$\omega' \cong \omega_0\left[1 \pm \frac{\beta}{\omega_0}\right] \tag{6}$$

which gives

$$\Delta\omega = \big(\omega_0 + \beta\big) - \big(\omega_0 - \beta\big) = 2\beta \tag{7}$$

We also can approximate ω_R for a lightly damped oscillator:

$$\omega_R = \sqrt{\omega_0^2 - 2\beta^2} \cong \omega_0 \tag{8}$$

Therefore, Q for a lightly damped oscillator becomes

$$Q \cong \frac{\omega_0}{2\beta} \cong \frac{\omega_0}{\Delta\omega} \tag{9}$$

3-24.   As requested, we use Equations (3.40), (3.57), and (3.60) with the given values to evaluate the complementary and particular solutions to the driven oscillator. The amplitude of the complementary function is constant as we vary ω, but the amplitude of the particular solution becomes larger as ω goes through the resonance near $0.96 \text{ rad} \cdot \text{s}^{-1}$, and decreases as ω is increased further. The plot closest to resonance here has $\omega/\omega_1 = 1.1$, which shows the least distortion due to transients. These figures are shown in figure (a). In figure (b), the $\omega/\omega_1 = 6$ plot from figure (a) is reproduced along with a new plot with $A_p = 20 \text{ m} \cdot \text{s}^{-2}$.

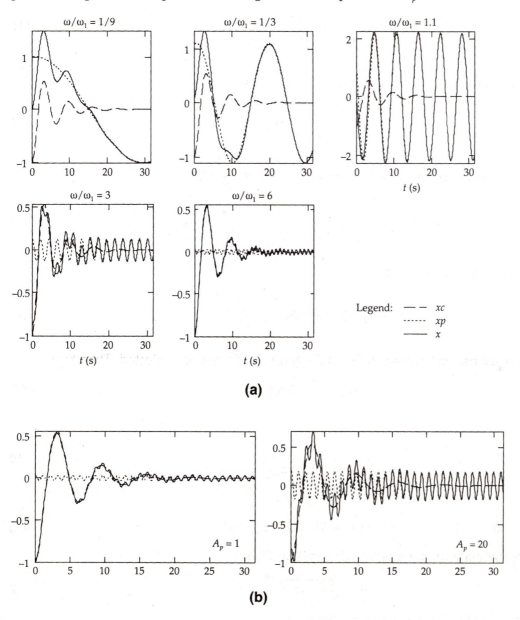

(a)

(b)

3-26. The equations of motion of this system are

$$m_1 \ddot{x}_1 = -kx_1 - b_1 (\dot{x}_1 - \dot{x}_2) + F \cos \omega t$$
$$m_2 \ddot{x}_2 = -b_2 \dot{x}_2 - b_1 (\dot{x}_2 - \dot{x}_1) \tag{1}$$

The electrical analog of this system can be constructed if we substitute in (1) the following equivalent quantities:

$$m_1 \rightarrow L_1; \quad k \rightarrow \frac{1}{C}; \quad b_1 \rightarrow R_1; \quad x \rightarrow q$$

$$m_2 \rightarrow L_2; \quad F \rightarrow \varepsilon_0; \quad b_2 \rightarrow R_2$$

Then the equations of the equivalent electrical circuit are given by

$$L_1 \ddot{q}_1 + R_1 (\dot{q}_1 - \dot{q}_2) + \frac{1}{C} q_1 = \varepsilon_0 \cos \omega t$$
$$L_2 \ddot{q}_2 + R_2 \dot{q}_2 + R_1 (\dot{q}_2 - \dot{q}_1) = 0 \tag{2}$$

Using the mathematical device of writing $\exp(i\omega t)$ instead of $\cos \omega t$ in (2), with the understanding that in the results only the real part is to be considered, and differentiating with respect to time, we have

$$L_1 \ddot{I}_1 + R_1 (\dot{I}_1 - \dot{I}_2) + \frac{I_1}{C} = i\omega \varepsilon_0 e^{i\omega t}$$
$$L_2 \ddot{I}_2 + R_2 (\dot{I}_2) + R_1 (\dot{I}_2 - \dot{I}_1) = 0 \tag{3}$$

Then, the equivalent electrical circuit is as shown in the figure:

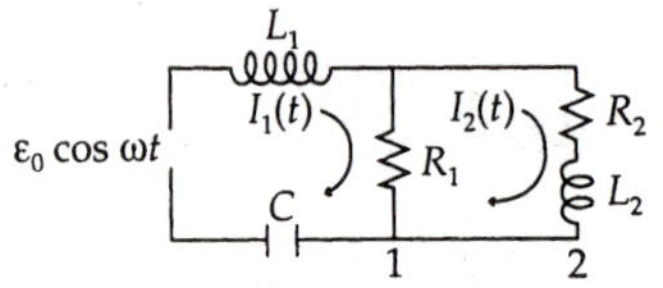

The impedance of the system Z is

$$Z = i\omega L_1 - i \frac{1}{\omega C} + Z_1 \tag{4}$$

where Z_1 is given by

$$\frac{1}{Z_1} = \frac{1}{R_1} + \frac{1}{R_2 + i\omega L_2} \tag{5}$$

Then,

$$Z_1 = \frac{R_1 \left[R_2 (R_2 + R_1) + \omega^2 L_2^2 + i\omega L_2 R_1 \right]}{(R_1 + R_2)^2 + \omega^2 L_2^2} \tag{6}$$

and substituting (6) into (4), we obtain

$$Z = \frac{R_1\left[R_2(R_2 + R_1) + \omega^2 L_2^2\right] + i\left[R_1\omega L_2 + \left(\omega L_1 - \dfrac{1}{\omega C}\right)\left((R_1 + R_2)^2 + \omega^2 L_2^2\right)\right]}{(R_1 + R_2)^2 + \omega^2 L_2^2} \tag{7}$$

3-29. In order to Fourier analyze a function of arbitrary period, say $\tau = 2P/\omega$ instead of $2\pi/\omega$, proportional change of scale is necessary. Analytically, such a change of scale can be represented by the substitution

$$x = \frac{\pi t}{P} \quad \text{or} \quad t = \frac{Px}{\pi} \tag{1}$$

for when $t = 0$, then $x = 0$, and when $t = \tau = 2P/\omega$, then $x = 2\pi/\omega$.

Thus, when the substitution $t = Px/\pi$ is made in a function $F(t)$ of period $2P/\omega'$, we obtain the function

$$F\left[\frac{Px}{\pi}\right] = f(x) \tag{2}$$

and this, *as a function of x*, has a period of $2\pi/\omega$. Now, $f(x)$ can, of course, be expanded according to the standard formula, Eq. (3.91):

$$f(x) = \frac{1}{2}a_0 + \sum_{n=1}^{\infty}\left(a_n \cos n\omega x + b_n \sin n\omega x\right) \tag{3}$$

where

$$\left.\begin{array}{l} a_n = \dfrac{\omega}{\pi}\displaystyle\int_0^{\frac{2\pi}{\omega}} f(x')\cos n\omega x'\, dx' \\[4mm] b_n = \dfrac{\omega}{\pi}\displaystyle\int_0^{\frac{2\pi}{\omega}} f(x')\sin n\omega x'\, dx' \end{array}\right\} \tag{4}$$

If, in the above expressions, we make the inverse substitutions

$$x = \frac{\pi t}{P} \quad \text{and} \quad dx = \frac{\pi}{P}dt \tag{5}$$

the expansion becomes

$$f\left[\frac{\pi t}{P}\right] = F\left[\frac{P}{\pi}\cdot\frac{\pi t}{P}\right] = F(t) = \frac{a_0}{2} + \sum_{n=1}^{\infty}\left[a_n \cos\left[\frac{n\omega\pi t}{P}\right] + b_n \sin\left[\frac{n\omega\pi t}{P}\right]\right] \tag{6}$$

and the coefficients in (4) become

$$a_n = \frac{\omega}{P} \int_0^{\frac{2P}{\omega}} F(t') \cos\left[\frac{n\omega\pi t'}{P}\right] dt'$$

$$b_n = \frac{\omega}{P} \int_0^{\frac{2P}{\omega}} F(t') \sin\left[\frac{n\omega\pi t'}{P}\right] dt' \qquad (7)$$

For the case corresponding to this problem, the period of $F(t)$ is $\frac{4\pi}{\omega}$, so that $P = 2\pi$. Then,

substituting into (7) and replacing the integral limits 0 and τ by the limits $-\frac{\tau}{2}$ and $+\frac{\tau}{2}$, we

obtain

$$a_n = \frac{\omega}{2\pi} \int_{-\frac{2\pi}{\omega}}^{\frac{2\pi}{\omega}} F(t') \cos\left[\frac{n\omega t'}{2}\right] dt'$$

$$b_n = \frac{\omega}{2\pi} \int_{-\frac{2\pi}{\omega}}^{\frac{2\pi}{\omega}} F(t') \sin\left[\frac{n\omega t'}{2}\right] dt' \qquad (8)$$

and substituting into (6), the expansion for $F(t)$ is

$$F(t) = \frac{a_0}{2} + \sum_{n=1}^{\infty} \left[a_n \cos\left[\frac{n\omega t}{2}\right] + b_n \sin\left[\frac{n\omega t}{2}\right] \right] \qquad (9)$$

Substituting $F(t)$ into (8) yields

$$a_n = \frac{\omega}{2\pi} \int_0^{\frac{2\pi}{\omega}} \sin\omega t' \cos\left[\frac{n\omega t'}{2}\right] dt'$$

$$b_n = \frac{\omega}{2\pi} \int_0^{\frac{2\pi}{\omega}} \sin\omega t' \sin\left[\frac{n\omega t'}{2}\right] dt' \qquad (10)$$

Evaluation of the integrals gives

$$b_2 = \frac{1}{2}; \ b_n = 0 \qquad \text{for } n \neq 2$$

$$a_0 = a_1 = 0 \qquad a_n(n \geq 2) = \begin{bmatrix} 0 & n \text{ even} \\ \dfrac{-4}{\pi(n^2 - 4)} & n \text{ odd} \end{bmatrix} \qquad (11)$$

and the resulting Fourier expansion is

$$F(t) = \frac{1}{2}\sin\omega t + \frac{4}{3\pi}\cos\frac{\omega t}{2} - \frac{4}{5\pi}\cos\frac{3\omega t}{2} - \frac{4}{21\pi}\cos\frac{5\omega t}{2} - \frac{4}{45\pi}\cos\frac{7\omega t}{2} + \dots \qquad (12)$$

3-32.

a) *Response to a Step Function:*

From Eq. (3.100) $H(t_0)$ is defined as

$$H(t_0) = \begin{bmatrix} 0, & t < t_0 \\ \\ a_1, & t > t_0 \end{bmatrix} \tag{1}$$

With initial conditions $x(t_0 = 0)$ and $\dot{x}(t_0 = 0)$, the general solution to Eq. (3.102) (equation of motion of a damped linear oscillator) is given by Eq. (3.105):

$$\begin{bmatrix} x(t) = \dfrac{a}{\omega_0^2}\left[1 - e^{-\beta(t-t_0)}\cos\omega_1(t-t_0) - \dfrac{\beta e^{-\beta(t-t_0)}}{\omega_1}\sin\omega_1(t-t_0)\right] \text{for } t > t_0 \\ \\ x(t) = 0 \hspace{7cm} \text{for } t < t_0 \end{bmatrix} \tag{2}$$

where $\omega_1 = \sqrt{\omega_0^2 - \beta^2}$.

For the case of overdamping, $\omega_0^2 < \beta^2$, and consequently $\omega_1 = i\sqrt{\beta^2 - \omega_0^2}$ is a pure imaginary number. Hence, $\cos\omega_1(t-t_0)$ and $\sin\omega_1(t-t_0)$ are no longer oscillatory functions; instead, they are transformed into hyperbolic functions. Thus, if we write $\omega_2 = \sqrt{\beta^2 - \omega_0^2}$ (where ω_2 is *real*),

$$\begin{bmatrix} \cos\omega_1(t-t_0) = \cos i\omega_2(t-t_0) = \cosh\omega_2(t-t_0) \\ \\ \sin\omega_1(t-t_0) = \sin i\omega_2(t-t_0) = i\sinh\omega_2(t-t_0) \end{bmatrix} \tag{3}$$

The response is given by [see Eq. (3.105)]

$$\begin{bmatrix} x(t) = \dfrac{a}{\omega_0^2}\left[1 - e^{-\beta(t-t_0)}\cosh\omega_2(t-t_0) - \dfrac{\beta e^{-\beta(t-t_0)}}{\omega_2}\sinh\omega_1(t-t_0)\right] \text{for } t > t_0 \\ \\ x(t) = 0 \hspace{7cm} \text{for } t < t_0 \end{bmatrix} \tag{4}$$

For simplicity, we choose $t_0 = 0$, and the solution becomes

$$\boxed{x(t) = \dfrac{H(0)}{\omega_0^2}\left[1 - e^{-\beta t}\cosh\omega_2 t - \dfrac{\beta e^{-\beta t}}{\omega_2}\sinh\omega_2 t\right]} \tag{5}$$

This response is shown in (a) below for the case $\beta = \sqrt{5}\,\omega_0$.

b) *Response to an Impulse Function* (in the limit $\tau \to 0$):

From Eq. (3.101) the impulse function $I(t_0, t_1)$ is defined as

$$I(t_0, t_1) = \begin{bmatrix} 0 & t < t_0 \\ a & t_0 < t < t_1 \\ 0 & t > t_1 \end{bmatrix} \tag{6}$$

For $t_1 - t_2 = \tau \to 0$ in such a way that $a\tau$ is constant $= b$, the response function is given by Eq. (3.110):

$$x(t) = \frac{b}{\omega_1} e^{-\beta(t-t_0)} \sin \omega_1 (t - t_0) \quad \text{for } t > t_0 \tag{7}$$

Again taking the "spike" to be at $t = 0$ for simplicity, we have

$$x(t) = \frac{b}{\omega_1} e^{-\beta t} \sin \omega_1 (t) \quad \text{for } t > 0 \tag{8}$$

For $\omega_1 = i\omega_2 = i\sqrt{\beta^2 - \omega_0^2}$ (overdamped oscillator), the solution is

$$\boxed{x(t) = \frac{b}{\omega_2} e^{-\beta t} \sinh \omega_2 t; \quad t > 0} \tag{9}$$

This response is shown in (b) below for the case $\beta = \sqrt{5}\,\omega_0$.

(a)

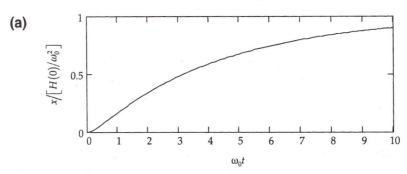

(b)

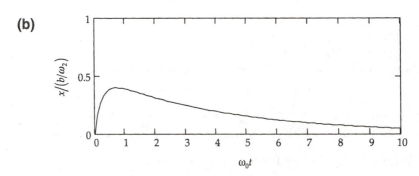

3-37. Any function $F(t)/m$ can be expanded in terms of step functions, as shown in the figure below where the curve is the sum of the various (positive and negative) step functions.

In general, we have

$$\ddot{x} + 2\beta\dot{x} + \omega_0^2 x = \sum_{n=-\infty}^{\infty} \frac{F_n(t)}{m}$$

$$= \sum_{n=-\infty}^{\infty} H_n(t) \tag{1}$$

where

$$H_n(t) = \begin{bmatrix} a_n(t) & t > t_n = n\tau \\ 0 & t < t_n = n\tau \end{bmatrix} \tag{2}$$

Then, since (1) is a *linear* equation, the solution to a superposition of functions of the form given by (2) is the superposition of the solutions for each of those functions.

According to Eq. (3.105), the solution for $H_n(t)$ for $t > t_n$ is

$$x_n(t) = \frac{a_n}{\omega_0^2}\left[1 - e^{-\beta(t-t_n)}\cos\omega_1(t-t_n) - \frac{\beta e^{-\beta(t-t_n)}}{\omega_1}\sin\omega_1(t-t_n)\right] \tag{3}$$

then, for

$$\frac{F(t)}{m} = \sum_{n=-\infty}^{\infty} H_n(t) \tag{4}$$

the solution is

$$x(t) = \frac{1}{\omega_0^2}\sum_{n=-\infty}^{\infty} H_n(t)\left[1 - e^{-\beta(t-t_n)}\cos\omega_1(t-t_n) - \frac{\beta e^{-\beta(t-t_n)}}{\omega_1}\sin\omega_1(t-t_n)\right]$$

$$= \sum_{n=-\infty}^{\infty} mH_n(t)G_n(t) = \sum_{n=-\infty}^{\infty} F_n(t)G_n(t) \tag{5}$$

where

$$G_n(t) = \begin{bmatrix} \dfrac{1}{m\omega_0^2}\left[1 - e^{-\beta(t-t_n)}\cos\omega_1(t-t_n) - \dfrac{\beta e^{-\beta(t-t_n)}}{\omega_1}\sin\omega_1(t-t_n)\right]; & t \geq t_n \\ \\ 0 & t < t_n \end{bmatrix} \tag{6}$$

or, comparing with (3)

$$G_n(t) = \begin{bmatrix} x_n(t)/ma_n, & t \geq t_n \\ 0 & t < t_n \end{bmatrix} \tag{7}$$

Therefore, the Green's function is the response to the unit step.

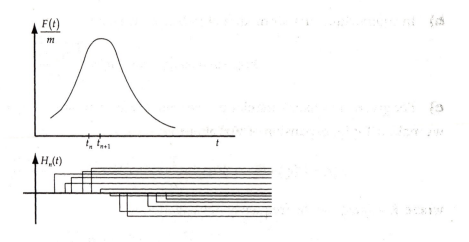

3-43.

a) Potential energy is the elastic energy:

$$U(r) = \frac{1}{2}k(r-a)^2,$$

where m is moving in a central force field. Then the effective potential is (see for example, Chapter 2 and Equation (8.14)):

$$U_{eff}(r) = U(r) + \frac{l^2}{2mr^2} = \frac{1}{2}k(r-a)^2 + \frac{l^2}{2mr^2}$$

where $l = mvr = m\omega r^2$ is the angular momentum of m and is a conserved quantity in this problem. The solid line below is $U_{eff}(r)$; at low values of r, the dashed line represents

$U(r) = \frac{1}{2}k(r-a)^2$, and the solid line is dominated by $\frac{l^2}{2mr^2}$. At large values of r,

$U_{eff}(r) \cong U(r) = \frac{1}{2}k(r-a)^2$.

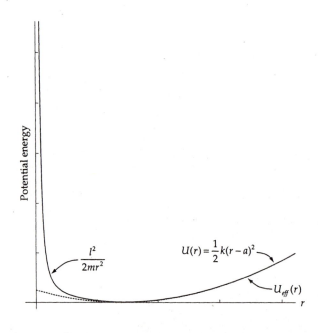

b) In equilibrium circular motion of radius r_0, we have

$$k(r_0 - a) = m\omega_0^2 r_0 \quad \Rightarrow \quad \omega_0 = \sqrt{\frac{k(r_0 - a)}{mr_0}}$$

c) For given (and fixed) angular momentum l, $V(r)$ is minimal at r_0, because $V'(r)\big|_{r=r_0} = 0$, so we make a Taylor expansion of $V(r)$ about r_0;

$$V(r) = V(r_0) + (r - r_0)V'(r_0) + \frac{1}{2}(r - r_0)^2 V''(r_0) + \ldots \approx \frac{3m\omega_0^2(r - r_0)^2}{2} = \frac{K(r - r_0)^2}{2}$$

where $K = 3m\omega_0^2$, so the frequency of oscillation is

$$\omega = \sqrt{\frac{K}{m}} = \sqrt{3}\,\omega_0 = \sqrt{\frac{3k(r_0 - a)}{mr_0}}$$

Nonlinear Oscillations and Chaos

4-2. Using the general procedure explained in Section 4.3, the phase diagram is constructed as follows:

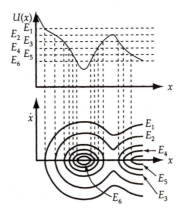

4-4. Differentiation of Rayleigh's equation above yields

$$\ddot{x} - \left(a - 3b\dot{x}^2\right)\ddot{x} + \omega_0^2 \dot{x} = 0 \qquad (1)$$

The substitution,

$$y = y_0 \sqrt{\frac{3b}{a}}\; \dot{x} \qquad (2)$$

implies that

$$\dot{x} = \sqrt{\frac{a}{3b}} \frac{y}{y_0}$$

$$\ddot{x} = \sqrt{\frac{a}{3b}} \frac{\dot{y}}{y_0} \qquad (3)$$

$$\dddot{x} = \sqrt{\frac{a}{3b}} \frac{\ddot{y}}{y_0}$$

When these expressions are substituted in (1), we find

$$\sqrt{\frac{a}{3b}} \frac{\ddot{y}}{y_0} - \sqrt{\frac{a}{3b}} \left[a - \frac{3ba}{b} \frac{\dot{y}^2}{y_0^2} \right] \frac{\dot{y}}{y_0} + \omega_0^2 \sqrt{\frac{a}{3b}} \frac{y}{y_0} = 0 \qquad (4)$$

Multiplying by $y_0 \sqrt{\dfrac{3b}{a}}$ and rearranging, we arrive at van der Pol's equation:

$$\boxed{\ddot{y} - \frac{a}{y_0^2} \left(y_0^2 - \dot{y}^2 \right) \dot{y} + \omega_0^2 y = 0} \qquad (5)$$

4-9.   The proposed force derives from a potential of the form

$$U(x) = \begin{bmatrix} \dfrac{1}{2} kx^2 & |x| < a \\ \dfrac{1}{2} (x + \delta) x - \delta a x^2 & |x| > a \end{bmatrix}$$

which is plotted in (a) below.

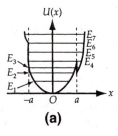

(a)

For small deviations from the equilibrium position ($x = 0$), the motion is just that of a harmonic oscillator.

For energies $E < E_6$, the particle cannot reach regions with $x < -a$, but it can reach regions of $x > a$ if $E > E_4$. For $E_2 < E < E_4$ the possibility exists that the particle can be trapped near $x = a$.

A phase diagram for the system is shown in (b) below.

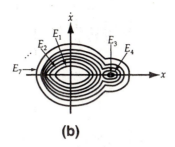

(b)

4-13.

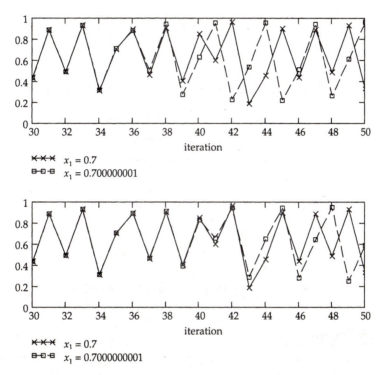

The plots are created by iteration on the initial values of (i) 0.7, (ii) 0.700000001, and (iii) 0.7000000001, using the equation

$$x_{n+1} = 2.5 \cdot x_n \left(1 - x_n^2\right) \tag{1}$$

A subset of the iterates from (i) and (ii) are plotted together, and clearly diverge by $n = 39$. The plot of (i) and (iii) clearly diverge by $n = 43$.

4-17.

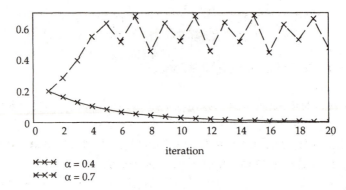

×—×—× $\alpha = 0.4$
×—×—× $\alpha = 0.7$

The first plot (with $\alpha = 0.4$) converges rather rapidly to zero, but the second (with $\alpha = 0.7$) does appear to be chaotic.

4-20.

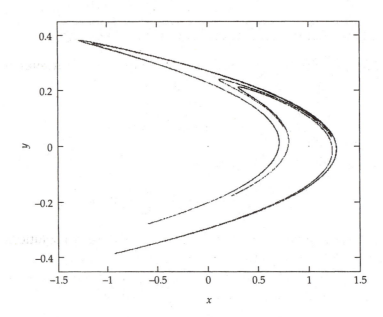

4-24.

a) The Van de Pol equation is

$$\frac{d^2x}{dt^2} + \omega_0^2 x = \mu(a^2 - x^2)\frac{dx}{dt}$$

Now look for solution in the form

$$x(t) = b \cos \omega_0 t + u(t) \tag{1}$$

we have

$$\frac{dx}{dt} = -b\omega_0 \sin \omega_0 t + \frac{du}{dt}$$

and

$$\frac{d^2x}{dt^2} = -b\omega_0^2 \cos \omega_0 t + \frac{d^2u}{dt^2}$$

Putting these into the Van de Pol equation, we obtain

$$\frac{d^2u(t)}{dt^2} + \omega_0 u(t) = -\mu\left\{b^2 \cos^2 \omega_0 t + u^2(t) + 2bu(t) \cos \omega_0 t - a^2\right\}\left\{-b\omega_0 \sin \omega_0 t + \frac{du(t)}{dt}\right\}$$

From this one can see that $u(t)$ is of order μ (i.e. $u \sim O(\mu)$), which is assumed to be small here. Keeping only terms up to order μ, the above equation reads

$$\frac{d^2u(t)}{dt^2} + \omega_0 u(t) = -\mu\left\{-b^3\omega_0 \sin \omega_0 t \cos^2 \omega_0 t + a^2 b\omega_0 \sin \omega_0 t\right\}$$

$$= -\mu b\omega_0\left\{\left(a^2 - \frac{b^2}{4}\right)\sin \omega_0 t - \frac{b^2}{4}\sin 3\omega_0 t\right\}$$

(where we have used the identity $4 \sin \omega_0 t \cos^2 \omega_0 t = \sin \omega_0 t + \sin 3\omega_0 t$)

This equation has 2 frequencies (ω_0 and $3\omega_0$), and is complicated. However, if $b = 2a$ then the term $\sin \omega_0 t$ disappears and the above equation becomes

$$\frac{d^2u(t)}{dt^2} + \omega_0 u(t) = \mu\omega_0 \frac{b^3}{4}\sin 3\omega_0 t$$

We let $b = 2a$, and the solution for this equation is

$$u(t) = -\frac{\mu b^3}{32\omega_0}\sin 3\omega_0 t = -\frac{\mu a^3}{4\omega_0}\sin 3\omega_0 t$$

So, finally putting this form of $u(t)$ into (1), we obtain one of the exact solutions of Van de Pol equation:

$$u(t) = 2a \cos \omega_0 t - \frac{\mu a^3}{4\omega_0}\sin 3\omega_0 t$$

b)   See phase diagram below. Since $\mu = 0.05$ is very small, then actually the second term in the expression of $u(t)$ is negligible, and the phase diagram is very close to a circle of radius $b = 2a = 2$.

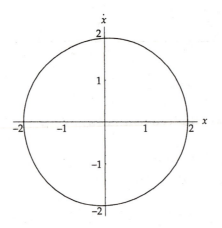

CHAPTER 5

Gravitation

5-2. Inside the sphere the gravitational potential satisfies

$$\nabla^2 \phi = 4\pi G\, \rho(r) \tag{1}$$

Since $\rho(r)$ is spherically symmetric, ϕ is also spherically symmetric. Thus,

$$\frac{1}{r^2}\frac{\partial}{\partial r}\left[r^2\frac{\partial \phi}{\partial r}\right] = 4\pi G\, \rho(r) \tag{2}$$

The field vector is independent of the radial distance. This fact implies

$$\frac{\partial \phi}{\partial r} = \text{constant} \equiv C \tag{3}$$

Therefore, (2) becomes

$$\frac{2C}{r} = 4\pi G\rho \tag{4}$$

or,

$$\boxed{\rho = \frac{C}{2\pi Gr}} \tag{5}$$

5-5. The equation of motion is

$$m\ddot{x} = -G\frac{Mm}{x^2} \tag{1}$$

Using conservation of energy, we find

$$\frac{1}{2}\dot{x}^2 - G\,M\frac{1}{x} = E = -G\,M\frac{1}{x_\infty} \tag{2}$$

$$\frac{dx}{dt} = -\sqrt{2GM\left[\frac{1}{x} - \frac{1}{x_\infty}\right]} \tag{3}$$

where x_∞ is some fixed large distance. Therefore, the time for the particle to travel from x_∞ to x is

$$t = -\int_{x_\infty}^{x} \frac{dx}{\sqrt{2GM\left[\frac{1}{x} - \frac{1}{x_\infty}\right]}} = -\frac{1}{\sqrt{GM}} \int_{x_\infty}^{x} \sqrt{\frac{x x_\infty}{2(x_\infty - x)}}\, dx$$

Making the change of variable, $x \to y^2$, and using Eq. (E.7), Appendix E, we obtain

$$t = \sqrt{\frac{x_\infty}{2GM}} \left[\sqrt{x(x_\infty - x)} - x_\infty \sin^{-1}\sqrt{\frac{x}{x_\infty}} \right]_{x_\infty}^{x} \tag{4}$$

If we set $x = 0$ and $x = x_\infty/2$ in (4), we can obtain the time for the particle to travel the total distance and the first half of the distance.

$$T_0 = \int_{x_\infty}^{0} dt = \frac{1}{\sqrt{GM}} \left[\frac{x_\infty}{2}\right]^{3/2} \tag{5}$$

$$T_{1/2} = \int_{x_\infty}^{x_\infty/2} dt = \frac{1}{\sqrt{GM}} \left[\frac{x_\infty}{2}\right]^{3/2} \left[1 + \frac{\pi}{2}\right] \tag{6}$$

Hence,

$$\frac{T_{1/2}}{T_0} = \frac{1 + \frac{\pi}{2}}{\pi}$$

Evaluating the expression,

$$\frac{T_{1/2}}{T_0} = 0.818 \tag{7}$$

or

$$\boxed{\frac{T_{1/2}}{T_0} \cong \frac{9}{11}} \tag{8}$$

5-10.

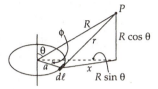

Using the relations

$$x = \sqrt{(R\sin\theta)^2 + a^2 - 2aR\sin\theta\cos\phi} \tag{1}$$

$$r = \sqrt{x^2 + R^2\cos^2\theta} = \sqrt{R^2 + a^2 - 2aR\sin\theta\cos\phi} \tag{2}$$

$$\rho_\ell = \frac{M}{2\pi a} \text{ (the linear mass density),} \tag{3}$$

the potential is expressed by

$$\Phi = -G \int \frac{\rho_\ell d\ell}{r} = \frac{-GM}{2\pi R} \int_0^{2\pi} \frac{d\phi}{\sqrt{1 - \left[2\dfrac{a}{R}\sin\theta\cos\phi - \dfrac{a^2}{R^2}\right]}} \tag{4}$$

If we expand the integrand and neglect terms of order $(a/R)^3$ and higher, we have

$$\left[1 - \left[2\frac{a}{R}\sin\theta\cos\phi - \frac{a^2}{R^2}\right]\right]^{-1/2} \cong 1 + \frac{a}{R}\sin\theta\cos\phi - \frac{1}{2}\frac{a^2}{R^2} + \frac{3}{2}\frac{a^2}{R^2}\sin^2\theta\cos^2\phi \tag{5}$$

Then, (4) becomes

$$\Phi \cong -\frac{GM}{2\pi R}\left[2\pi - \frac{1}{2}\frac{a^2}{R^2}2\pi + \frac{3}{2}\frac{a^2}{R^2}\pi\sin^2\theta\right]$$

Thus,

$$\boxed{\Phi(R) \cong -\frac{GM}{R}\left[1 - \frac{1}{2}\frac{a^2}{R^2}\left[1 - \frac{3}{2}\sin^2\theta\right]\right]} \tag{6}$$

5-16.

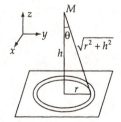

For points external to the sphere, we may consider the sphere to be a point mass of mass M. Put the sheet in the x-y plane.

Consider force on M due to the sheet. By symmetry, $F_x = F_y = 0$

$$F_z = \int dF_z = \int_{r=0}^{\infty} \frac{GMdm}{\left(r^2 + h^2\right)} \cos\theta$$

With $dm = \rho_s 2\pi r dr$ and $\cos\theta = \dfrac{h}{\sqrt{r^2 + h^2}}$

we have

$$F_z = 2\pi\rho_s GMh \int_{r=0}^{\infty} \frac{rdr}{\left(r^2 + h^2\right)^{3/2}}$$

$$F_z = -2\pi\rho_s GMh \left[\frac{1}{\left(r^2 + h^2\right)^{1/2}} \right]_0^{\infty}$$

$$F_z = 2\pi\rho_s GM$$

> The sphere attracts the sheet in the z-direction
> with a force of magnitude $2\pi\rho_s GM$

5-18. From Equation (5.55), we have with the appropriate substitutions

$$\frac{h_{\text{moon}}}{h_{\text{sun}}} = \frac{\dfrac{3GM_m r^2}{2gD^3}}{\dfrac{3GM_s r^2}{2gR_{es}^3}} = \frac{M_m}{M_s}\left[\frac{R_{es}}{D}\right]^3 \tag{1}$$

Substitution of the known values gives

$$\frac{h_{\text{moon}}}{h_{\text{sun}}} = \frac{7.350 \times 10^{22} \text{ kg}}{1.993 \times 10^{30} \text{ kg}} \left[\frac{1.495 \times 10^{11} \text{ m}}{3.84 \times 10^8 \text{ m}}\right]^3 \approx 2.2 \tag{2}$$

CHAPTER 6

Some Methods in the Calculus of Variations

6-3. The element of distance in three-dimensional space is

$$dS = \sqrt{dx^2 + dy^2 + dz^2} \tag{1}$$

Suppose x, y, z depends on the parameter t and that the end points are expressed by $\left(x_1(t_1), y_1(t_1), z_1(t_1)\right)$, $\left(x_2(t_2), y_2(t_2), z_2(t_2)\right)$. Then the total distance is

$$S = \int_{t_1}^{t_2} \sqrt{\left[\frac{dx}{dt}\right]^2 + \left[\frac{dy}{dt}\right]^2 + \left[\frac{dz}{dt}\right]^2} \, dt \tag{2}$$

The function f is identified as

$$f = \sqrt{\dot{x}^2 + \dot{y}^2 + \dot{z}^2} \tag{3}$$

Since $\dfrac{\partial f}{\partial x} = \dfrac{\partial f}{\partial y} = \dfrac{\partial f}{\partial z} = 0$, the Euler equations become

$$\left.\begin{array}{c} \dfrac{d}{dt}\dfrac{\partial f}{\partial \dot{x}} = 0 \\[2mm] \dfrac{d}{dt}\dfrac{\partial f}{\partial \dot{y}} = 0 \\[2mm] \dfrac{d}{dt}\dfrac{\partial f}{\partial \dot{z}} = 0 \end{array}\right] \tag{4}$$

from which we have

$$\frac{\dot{x}}{\sqrt{\dot{x}^2 + \dot{y}^2 + \dot{z}^2}} = \text{constant} \equiv C_1$$

$$\frac{\dot{y}}{\sqrt{\dot{x}^2 + \dot{y}^2 + \dot{z}^2}} = \text{constant} \equiv C_2 \qquad (5)$$

$$\frac{\dot{z}}{\sqrt{\dot{x}^2 + \dot{y}^2 + \dot{z}^2}} = \text{constant} \equiv C_3$$

From the combination of these equations, we have

$$\frac{\dot{x}}{C_1} = \frac{\dot{y}}{C_2}$$

$$\frac{\dot{y}}{C_2} = \frac{\dot{z}}{C_3} \qquad (6)$$

If we integrate (6) from t_1 to the arbitrary t, we have

$$\frac{x - x_1}{C_1} = \frac{y - y_1}{C_2}$$

$$\frac{y - y_1}{C_2} = \frac{z - z_1}{C_3} \qquad (7)$$

On the other hand, the integration of (6) from t_1 to t_2 gives

$$\frac{x_2 - x_1}{C_1} = \frac{y_2 - y_1}{C_2}$$

$$\frac{y_2 - y_1}{C_2} = \frac{z_2 - z_1}{C_3} \qquad (8)$$

from which we find the constants C_1, C_2, and C_3. Substituting these constants into (7), we find

$$\boxed{\frac{x - x_1}{x_2 - x_1} = \frac{y - y_1}{y_2 - y_1} = \frac{z - z_1}{z_2 - z_1}} \qquad (9)$$

This is the equation expressing a straight line in three-dimensional space passing through the two points (x_1, y_1, z_1), (x_2, y_2, z_2).

6-7.

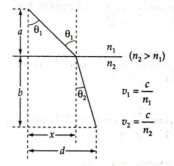

The time to travel the path shown is (cf. Example 6.2)

$$t = \int \frac{ds}{v} = \int \frac{\sqrt{1+y'^2}}{v} \, dx \qquad (1)$$

Although we have $v = v(y)$, we only have $dv/dy \neq 0$ when $y = 0$. The Euler equation tells us

$$\frac{d}{dx}\left[\frac{y'}{v\sqrt{1+y'^2}} \right] = 0 \qquad (2)$$

Now use $v = c/n$ and $y' = -\tan\theta$ to obtain

$$n \sin\theta = \text{const.} \qquad (3)$$

This proves the assertion. Alternatively, Fermat's principle can be proven by the method introduced in the solution of Problem 6-8.

6-10. This problem lends itself to the method of solution suggested in the solution of Problem 6-8. The volume of a right cylinder is given by

$$V = \pi R^2 H \qquad (1)$$

The total surface area A of the cylinder is given by

$$A = A_{\text{bases}} + A_{\text{side}} = 2\pi R^2 + 2\pi RH = 2\pi R(R + H) \qquad (2)$$

We wish A to be a minimum. (1) is the constraint condition, and the other equations are

$$\left.\begin{array}{c} \dfrac{\partial A}{\partial R} + \lambda \dfrac{\partial g}{\partial R} = 0 \\[2mm] \dfrac{\partial A}{\partial H} + \lambda \dfrac{\partial g}{\partial H} = 0 \end{array}\right\} \qquad (3)$$

where $g = V - \pi R^2 H = 0$.

The solution of these equations is

$$\boxed{R = \frac{1}{2}H} \qquad (4)$$

6-14. It is more convenient to work with cylindrical coordinates (r, ϕ, z) in this problem. The constraint here is $z = 1 - r$, then $dz = -dr$

$$ds^2 = dr^2 + r^2 d\phi^2 + dz^2 = 2\left(dr^2 + r^2 d\beta^2\right)$$

where we have introduced a new angular coordinate $\beta = \dfrac{\phi}{\sqrt{2}}$

In this form of ds^2, we clearly see that the space is 2-dimensional Euclidean flat, so the shortest line connecting two given points is a straight line given by:

$$r = \frac{r_0}{\cos(\beta - \beta_0)} = \frac{r_0}{\cos\left(\dfrac{\phi - \phi_0}{\sqrt{2}}\right)}$$

this line passes through the endpoints $(r = 1, \ \phi = \pm\dfrac{\pi}{2})$, then we can determine unambiguously the shortest path equation

$$r(\phi) = \frac{\cos\dfrac{\pi}{2\sqrt{2}}}{\cos\left(\dfrac{\phi}{\sqrt{2}}\right)} \quad \text{and} \quad z = 1 - r$$

Accordingly, the shortest connecting length is

$$l = \int_{-\pi/2}^{\pi/2} d\phi \sqrt{2\left(\frac{dr}{d\phi}\right)^2 + r^2} = 2\sqrt{2}\,\sin\frac{\pi}{2\sqrt{2}}$$

Hamilton's Principle— Lagrangian and Hamiltonian Dynamics

7-3.

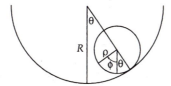

If we take angles θ and ϕ as our generalized coordinates, the kinetic energy and the potential energy of the system are

$$T = \frac{1}{2} m \left[(R - \rho) \dot{\theta} \right]^2 + \frac{1}{2} I \dot{\phi}^2 \tag{1}$$

$$U = \left[R - (R - \rho) \cos \theta \right] mg \tag{2}$$

where m is the mass of the sphere and where $U = 0$ at the lowest position of the sphere. I is the moment of inertia of sphere with respect to any diameter. Since $I = (2/5) m\rho^2$, the Lagrangian becomes

$$L = T - U = \frac{1}{2} m (R - \rho)^2 \dot{\theta}^2 + \frac{1}{5} m\rho^2 \dot{\phi}^2 - \left[R - (R - \rho) \cos \theta \right] mg \tag{3}$$

When the sphere is at its lowest position, the points A and B coincide. The condition $A0 = B0$ gives the equation of constraint:

$$f(\theta, \phi) = (R - \rho)\theta - \rho\phi = 0 \tag{4}$$

Therefore, we have two Lagrange's equations with one undetermined multiplier:

$$\left. \begin{array}{l} \dfrac{\partial L}{\partial \theta} - \dfrac{d}{dt} \left[\dfrac{\partial L}{\partial \dot{\theta}} \right] + \lambda \dfrac{\partial f}{\partial \theta} = 0 \\[3mm] \dfrac{\partial L}{\partial \phi} - \dfrac{d}{dt} \left[\dfrac{\partial L}{\partial \dot{\phi}} \right] + \lambda \dfrac{\partial f}{\partial \phi} = 0 \end{array} \right\} \tag{5}$$

After substituting (3) and $\partial f/\partial\theta = R - \rho$ and $\partial f/\partial\phi = -\rho$ into (5), we find

$$-(R-\rho)mg\sin\theta - m(R-\rho)^2\ddot{\theta} + \lambda(R-\rho) = 0 \tag{6}$$

$$-\frac{2}{5}m\rho^2\ddot{\phi} - \lambda\rho = 0 \tag{7}$$

From (7) we find λ:

$$\lambda = -\frac{2}{5}m\rho\ddot{\phi} \tag{8}$$

or, if we use (4), we have

$$\lambda = -\frac{2}{5}m(R-\rho)\ddot{\theta} \tag{9}$$

Substituting (9) into (6), we find the equation of motion with respect to θ:

$$\ddot{\theta} = -\omega^2\sin\theta \tag{10}$$

where ω is the frequency of small oscillations, defined by

$$\omega = \sqrt{\frac{5g}{7(R-\rho)}} \tag{11}$$

7-6.

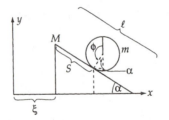

Let us choose ξ, S as our generalized coordinates. The x,y coordinates of the center of the hoop are expressed by

$$\left.\begin{aligned} x &= \xi + S\cos\alpha + r\sin\alpha \\ y &= r\cos\alpha + (\ell - S)\sin\alpha \end{aligned}\right\} \tag{1}$$

Therefore, the kinetic energy of the hoop is

$$\begin{aligned} T_{\text{hoop}} &= \frac{1}{2}m\left(\dot{x}^2 + \dot{y}^2\right) + \frac{1}{2}I\dot{\phi}^2 \\ &= \frac{1}{2}m\left[\left(\dot{\xi} + \dot{S}\cos\alpha\right)^2 + \left(-\dot{S}\sin\alpha\right)^2\right] + \frac{1}{2}I\dot{\phi}^2 \end{aligned} \tag{2}$$

Using $I = mr^2$ and $S = r\phi$, (2) becomes

$$T_{\text{hoop}} = \frac{1}{2} m \left[2\dot{S}^2 + \dot{\xi}^2 + 2\dot{\xi}\dot{S} \cos \alpha \right] \tag{3}$$

In order to find the total kinetic energy, we need to add the kinetic energy of the translational motion of the plane along the x-axis which is

$$T_{\text{plane}} = \frac{1}{2} M \dot{\xi}^2 \tag{4}$$

Therefore, the total kinetic energy becomes

$$T = m\dot{S}^2 + \frac{1}{2}(m + M)\dot{\xi}^2 + m\dot{\xi}\dot{S} \cos \alpha \tag{5}$$

The potential energy is

$$U = mgy = mg\left[r \cos \alpha + (\ell - S) \sin \alpha \right] \tag{6}$$

Hence, the Lagrangian is

$$l = m\dot{S}^2 + \frac{1}{2}(m + M)\dot{\xi}^2 + m\dot{\xi}\dot{S} \cos \alpha - mg\left[r \cos \alpha + (\ell - S) \sin \alpha \right] \tag{7}$$

from which the Lagrange equations for ξ and S are easily found to be

$$\boxed{2m\ddot{S} + m\ddot{\xi} \cos \alpha - mg \sin \alpha = 0} \tag{8}$$

$$\boxed{(m + M)\ddot{\xi} + m\ddot{S} \cos \alpha = 0} \tag{9}$$

or, if we rewrite these equations in the form of uncoupled equations by substituting for $\ddot{\xi}$ and $\ddot{S}$, we have

$$\boxed{\begin{array}{c} \left[2 - \dfrac{m \cos^2 \alpha}{m + M} \right] \ddot{S} - g \sin \alpha = 0 \\[4mm] \ddot{\xi} = -\dfrac{mg \sin \alpha \cos \alpha}{2(m + M) - m \cos^2 \alpha} \end{array}} \tag{10}$$

Now, we can rewrite (9) as

$$\frac{d}{dt}\left[(m + M)\dot{\xi} + m\dot{S} \cos \alpha \right] = 0 \tag{11}$$

where we can interpret $(m + M)\dot{\xi}$ as the x component of the linear momentum of the total system and $m\dot{S} \cos \alpha$ as the x component of the linear momentum of the hoop with respect to the plane. Therefore, (11) means that the x component of the total linear momentum is a constant of motion. This is the expected result because no external force is applied along the x-axis.

7-10.

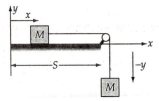

Let the length of the string be ℓ so that

$$(S-x)-y=\ell \tag{1}$$

Then,

$$\dot{x}=-\dot{y} \tag{2}$$

a) The Lagrangian of the system is

$$L=\frac{1}{2}M\dot{x}^2+\frac{1}{2}M\dot{y}^2-Mgy=M\dot{y}^2-Mgy \tag{3}$$

Therefore, Lagrange's equation for y is

$$\frac{d}{dt}\frac{\partial L}{\partial \dot{y}}-\frac{\partial L}{\partial y}=2M\ddot{y}+Mg=0 \tag{4}$$

from which

$$\ddot{y}=-\frac{g}{2} \tag{5}$$

Then, the general solution for y becomes

$$y(t)=-\frac{g}{4}t^2+C_1t+C_2 \tag{6}$$

If we assign the initial conditions $y(t=0)=0$ and $\dot{y}(t=0)=0$, we find

$$\boxed{y(t)=-\frac{g}{4}t^2} \tag{7}$$

b) If the string has a mass m, we must consider its kinetic energy and potential energy. These are

$$T_{string}=\frac{1}{2}m\dot{y}^2 \tag{8}$$

$$U_{string}=-\frac{m}{\ell}yg\frac{y}{2}=-\frac{mg}{2\ell}y^2 \tag{9}$$

Adding (8) and (9) to (3), the total Lagrangian becomes

$$L=M\dot{y}^2-Mgy+\frac{1}{2}m\dot{y}^2+\frac{mg}{2\ell}y^2 \tag{10}$$

Therefore, Lagrange's equation for y now becomes

$$(2M+m)\ddot{y} - \frac{mg}{\ell}y + Mg = 0 \tag{11}$$

In order to solve (11), we arrange this equation into the form

$$(2M+m)\ddot{y} = \frac{mg}{\ell}\left[y - \frac{M\ell}{m}\right] \tag{12}$$

Since $\frac{d^2}{dt^2}\left[y - \frac{M\ell}{m}\right] = \frac{d^2}{dt^2}y$, (12) is equivalent to

$$\frac{d^2}{dt^2}\left[y - \frac{M\ell}{m}\right] = \frac{mg}{\ell(2M+m)}\left[y - \frac{M\ell}{m}\right] \tag{13}$$

which is solved to give

$$y - \frac{M\ell}{m} = Ae^{\gamma t} + Be^{-\gamma t} \tag{14}$$

where

$$\gamma = \sqrt{\frac{mg}{\ell(2M+m)}} \tag{15}$$

If we assign the initial condition $y(t=0) = 0$; $\dot{y}(t=0) = 0$, we have

$$A = +B = -\frac{M\ell}{2m}$$

Then,

$$\boxed{y(t) = \frac{M\ell}{m}(1 - \cosh \gamma t)} \tag{16}$$

7-13.

a)

$$x = \frac{1}{2}at^2 - b\sin\theta$$

$$y = -b\cos\theta$$

$$\dot{x} = at - b\dot{\theta}\cos\theta$$

$$\dot{y} = b\dot{\theta}\sin\theta$$

$$L = \frac{1}{2} m\left(\dot{x}^2 + \dot{y}^2\right) - mgy$$

$$= \frac{1}{2} m\left(a^2 t^2 - 2at\, b\dot{\theta} \cos\theta + b^2\dot{\theta}^2\right) + mgb\cos\theta$$

$$\frac{d}{dt}\frac{\partial L}{\partial \dot{\theta}} = \frac{\partial L}{\partial \theta} \quad \text{gives}$$

$$\frac{d}{dt}\left[-mat\, b\cos\theta + mb^2\dot{\theta}\right] = mat\, b\ddot{\theta}\sin\theta - mgb\sin\theta$$

This gives the equation of motion

$$\boxed{\ddot{\theta} + \frac{g}{b}\sin\theta - \frac{a}{b}\cos\theta = 0}$$

b) To find the period for small oscillations, we must expand $\sin\theta$ and $\cos\theta$ about the equilibrium point θ_0. We find θ_0 by setting $\ddot{\theta} = 0$. For equilibrium,

$$g\sin\theta_0 = a\cos\theta_0$$

or

$$\tan\theta_0 = \frac{a}{g}$$

Using the first two terms in a Taylor series expansion for $\sin\theta$ and $\cos\theta$ gives

$$f(\theta) \simeq f(\theta_0) + f'(\theta)\big]_{\theta=\theta_0}(\theta - \theta_0)$$

$$\sin\theta \simeq \sin\theta_0 + (\theta - \theta_0)\cos\theta_0$$

$$\cos\theta \simeq \cos\theta_0 - (\theta - \theta_0)\sin\theta_0$$

$$\tan\theta_0 = \frac{a}{g} \quad \text{implies} \quad \sin\theta_0 = \frac{a}{\sqrt{a^2 + g^2}},$$

$$\cos\theta_0 = \frac{g}{\sqrt{a^2 + g^2}}$$

Thus

$$\sin\theta \simeq \frac{1}{\sqrt{a^2 + g^2}}\left(a + g\theta - g\theta_0\right)$$

$$\cos\theta \simeq \frac{1}{\sqrt{a^2 + g^2}}\left(g - a\theta + a\theta_0\right)$$

Substituting into the equation of motion gives

$$0 = \ddot{\theta} + \frac{g}{b\sqrt{a^2 + g^2}}(a + g\theta - g\theta_0) - \frac{a}{b\sqrt{a^2 + g^2}}(g - a\theta + a\theta_0)$$

This reduces to

$$\ddot{\theta} + \frac{\sqrt{g^2 + a^2}}{b}\theta = \frac{\sqrt{g^2 + a^2}}{b}\theta_0$$

The solution to this inhomogeneous differential equation is

$$\theta = \theta_0 + A\cos\omega\theta + B\sin\omega\theta$$

where

$$\omega = \frac{\left(g^2 + a^2\right)^{1/4}}{b^{1/2}}$$

Thus

$$\boxed{T = \frac{2\pi}{\omega} = \frac{2\pi\, b^{1/2}}{\left(g^2 + a^2\right)^{1/4}}}$$

7-16.

For mass m:

$$x = a\sin\omega t + b\sin\theta$$

$$y = -b\cos\theta$$

$$\dot{x} = a\omega\cos\omega t + b\dot{\theta}\cos\theta$$

$$\dot{y} = b\dot{\theta}\sin\theta$$

Substitute into

$$T = \frac{1}{2}m\left(\dot{x}^2 + \dot{y}^2\right)$$

$$U = mgy$$

and the result is

$$L = T - U = \frac{1}{2}m\left(a^2\omega^2\cos^2\omega t + 2ab\omega\dot{\theta}\cos\omega t\cos\theta + b^2\dot{\theta}^2\right) + mgb\cos\theta$$

Lagrange's equation for θ gives

$$\frac{d}{dt}\left(mab\omega\cos\omega t\cos\theta + mb^2\dot\theta\right) = -mab w\dot\theta\cos\omega t\sin\theta - mgb\sin\theta$$

$$-ab\omega^2\sin\omega t\cos\theta - ab\omega\dot\theta\cos\omega t\sin\theta + b^2\ddot\theta = -ab\omega\dot\theta\cos\omega t\sin\theta - gb\sin\theta$$

or

$$\boxed{\ddot\theta + \frac{g}{b}\sin\theta - \frac{a}{b}\omega^2\sin\omega t\cos\theta = 0}$$

7-20. The x-y plane is horizontal, and A, B, C are the fixed points lying in a plane above the hoop. The hoop rotates about the vertical through its center.

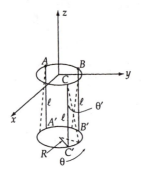

The kinetic energy of the system is given by

$$T = \frac{1}{2}I\omega^2 + \frac{1}{2}M\dot z^2 = \frac{MR^2}{2}\dot\theta^2 + \frac{1}{2}M\left[\frac{\partial z}{\partial\theta}\right]^2\dot\theta^2 \tag{1}$$

For small θ, the second term can be neglected since $\left.\left(\partial z/\partial\theta\right)\right|_{\theta=0} = 0$

The potential energy is given by

$$U = Mgz \tag{2}$$

where we take $U = 0$ at $z = -\ell$.

Since the system has only one degree of freedom we can write z in terms of θ. When $\theta = 0$, $z = -\ell$. When the hoop is rotated thorough an angle θ, then

$$z^2 = \ell^2 - \left(R - R\cos\theta\right)^2 - \left(R\sin\theta\right)^2 \tag{3}$$

so that

$$z = -\left[\ell^2 + 2R^2\left(\cos\theta - 1\right)\right]^{1/2} \tag{4}$$

and the potential energy is given by

$$U = -Mg\left[\ell^2 + 2R^2\left(\cos\theta - 1\right)\right]^{1/2} \tag{5}$$

for small θ, $\cos\theta - 1 \cong -\theta^2/2$; then,

$$U \cong -Mg\ell \left[1 - \frac{R^2\theta^2}{\ell^2} \right]^{1/2}$$

$$\cong -Mg\ell \left[1 - \frac{R^2\theta^2}{2\ell^2} \right] \tag{6}$$

From (1) and (6), the Lagrangian is

$$L = T - U = \frac{1}{2} MR^2\dot{\theta}^2 + Mg\ell \left[1 - \frac{R^2\theta^2}{2\ell^2} \right], \tag{7}$$

for small θ. The Lagrange equation for θ gives

$$\ddot{\theta} + \frac{g}{\ell}\theta = 0 \tag{8}$$

where

$$\boxed{\omega = \sqrt{\frac{g}{\ell}}} \tag{9}$$

which is the frequency of small rotational oscillations about the vertical through the center of the hoop and is the same as that for a simple pendulum of length ℓ.

7-23. The Hamiltonian function can be written as [see Eq. (7.153)]

$$H = \sum_j p_j \dot{q}_j - L \tag{1}$$

For a particle which moves freely in a conservative field with potential U, the Lagrangian in rectangular coordinates is

$$L = \frac{1}{2} m\left(\dot{x}^2 + \dot{y}^2 + \dot{z}^2 \right) - U$$

and the linear momentum components in rectangular coordinates are

$$\left. \begin{array}{l} p_x = \dfrac{\partial L}{\partial \dot{x}} = m\dot{x} \\[2mm] p_y = m\dot{y} \\[2mm] p_z = m\dot{z} \end{array} \right\} \tag{2}$$

$$H = \left[m\dot{x}^2 + m\dot{y}^2 + m\dot{z}^2 \right] - \left[\frac{1}{2} m\left(\dot{x}^2 + \dot{y}^2 + \dot{z}^2 \right) - U \right]$$

$$= \frac{1}{2} m\left(\dot{x}^2 + \dot{y}^2 + \dot{z}^2 \right) + U = \frac{1}{2m}\left(p_x^2 + p_y^2 + p_z^2 \right) \tag{3}$$

which is just the total energy of the particle. The canonical equations are [from Eqs. (7.160) and (7.161)]

$$
\begin{aligned}
\dot{p}_x &= m\ddot{x} = -\frac{\partial U}{\partial x} = F_x \\
\dot{p}_y &= m\ddot{y} = -\frac{\partial U}{\partial y} = F_y \\
\dot{p}_z &= m\ddot{z} = -\frac{\partial U}{\partial z} = F_z
\end{aligned}
\tag{4}
$$

These are simply Newton's equations.

7-30.

a) From the definition of a total derivative, we can write

$$
\frac{dg}{dt} = \frac{\partial g}{\partial t} + \sum_k \left[\frac{\partial g}{\partial q_k} \frac{\partial q_k}{\partial t} + \frac{\partial g}{\partial p_k} \frac{\partial p_k}{\partial t} \right]
\tag{1}
$$

Using the canonical equations

$$
\begin{aligned}
\frac{\partial q_k}{\partial t} &= \dot{q}_k = \frac{\partial H}{\partial p_k} \\[2mm]
\frac{\partial p_k}{\partial t} &= \dot{p}_k = -\frac{\partial H}{\partial q_k}
\end{aligned}
\tag{2}
$$

we can write (1) as

$$
\frac{dg}{dt} = \frac{\partial g}{\partial t} + \sum_k \left[\frac{\partial g}{\partial q_k} \frac{\partial H}{\partial p_k} - \frac{\partial g}{\partial p_k} \frac{\partial H}{\partial q_k} \right]
\tag{3}
$$

or

$$
\frac{dg}{dt} = \frac{\partial g}{\partial t} + [g, H]
\tag{4}
$$

b)
$$
\dot{q}_j = \frac{\partial q_j}{\partial t} = \frac{\partial H}{\partial p_j}
\tag{5}
$$

According to the definition of the Poisson brackets,

$$
[q_j, H] = \sum_k \left[\frac{\partial q_j}{\partial q_k} \frac{\partial H}{\partial p_k} - \frac{\partial g_j}{\partial p_k} \frac{\partial H}{\partial q_k} \right]
\tag{6}
$$

but

$$
\frac{\partial q_j}{\partial q_k} = \delta_{jk} \text{ and } \frac{\partial q_j}{\partial p_k} = 0 \text{ for any } j,k
\tag{7}
$$

then (6) can be expressed as

$$\boxed{\left[q_j, H\right] = \frac{\partial H}{\partial p_j} = \dot{q}_j}$$

(8)

In the same way, from the canonical equations,

$$\dot{p}_j = -\frac{\partial H}{\partial q_j}$$

(9)

so that

$$\left[p_j, H\right] = \sum_k \left[\frac{\partial p_j}{\partial q_k}\frac{\partial H}{\partial p_k} - \frac{\partial p_j}{\partial p_k}\frac{\partial H}{\partial q_k}\right]$$

(10)

but

$$\frac{\partial p_j}{\partial p_k} = \delta_{jk} \text{ and } \frac{\partial p_j}{\partial q_k} = 0 \text{ for any } j,k$$

(11)

then,

$$\boxed{\dot{p}_j = -\frac{\partial H}{\partial q_j} = \left[p_j, H\right]}$$

(12)

c)

$$\left[p_k, p_j\right] = \sum_\ell \left[\frac{\partial p_k}{\partial q_\ell}\frac{\partial p_j}{\partial p_\ell} - \frac{\partial p_k}{\partial p_\ell}\frac{\partial p_j}{\partial q_\ell}\right]$$

(13)

since,

$$\frac{\partial p_k}{\partial q_\ell} = 0 \text{ for any } k,\ell$$

(14)

the right-hand side of (13) vanishes, and

$$\boxed{\left[p_k, p_j\right] = 0}$$

(15)

In the same way,

$$\left[q_k, q_j\right] = \sum_\ell \left[\frac{\partial q_k}{\partial q_\ell}\frac{\partial q_j}{\partial p_\ell} - \frac{\partial q_k}{\partial p_\ell}\frac{\partial q_j}{\partial q_\ell}\right]$$

(16)

since

$$\frac{\partial q_j}{\partial p_\ell} = 0 \text{ for any } j,\ell$$

(17)

the right-hand side of (16) vanishes and

$$\boxed{\left[q_k, q_j\right] = 0}$$

(18)

d)

$$\left[q_k, p_j\right] = \sum_\ell \left[\frac{\partial q_k}{\partial q_\ell}\frac{\partial p_j}{\partial p_\ell} - \frac{\partial q_k}{\partial p_\ell}\frac{\partial p_j}{\partial q_\ell}\right]$$

$$= \sum_\ell \delta_{k\ell}\,\delta_{j\ell} \tag{19}$$

or,

$$\boxed{\left[q_k, p_j\right] = \delta_{kj}} \tag{20}$$

e) Let $g(p_k, q_k)$ be a quantity that does not depend explicitly on the time. If $g(p_k, q_k)$ commutes with the Hamiltonian, i.e., if

$$[g, H] = 0 \tag{21}$$

then, according to the result in a) above,

$$\boxed{\frac{dg}{dt} = 0} \tag{22}$$

and g is a constant of motion.

7-32. The Lagrangian for this case is

$$L = T - U = \frac{1}{2}m\left(\dot{r}^2 + r^2\dot{\theta}^2 + r^2\sin^2\theta\,\dot{\phi}^2\right) + \frac{k}{r} \tag{1}$$

where spherical coordinates have been used due to the symmetry of U.

The generalized coordinates are r, θ, and ϕ, and the generalized momenta are

$$p_r = \frac{\partial L}{\partial \dot{r}} = m\dot{r} \tag{2}$$

$$p_\theta = \frac{\partial L}{\partial \dot{\theta}} = mr^2\dot{\theta} \tag{3}$$

$$p_\phi = \frac{\partial L}{\partial \dot{\phi}} = mr^2\dot{\phi}\sin^2\theta \tag{4}$$

The Hamiltonian can be constructed as in Eq. (7.155):

$$H = p_r\dot{r} + p_\theta\dot{\theta} + p_\phi\dot{\phi} - L$$

$$= \frac{1}{2}m\left(\dot{r}^2 + r^2\dot{\theta}^2 + r^2\dot{\phi}^2\sin^2\theta\right) - \frac{k}{r}$$

$$= \frac{1}{2}\left[\frac{p_r^2}{m} + \frac{p_\theta^2}{mr^2} + \frac{p_\phi^2}{mr^2\sin^2\theta}\right] - \frac{k}{r} \tag{5}$$

Eqs. (7.160) applied to H as given in (5) reproduce equations (2), (3), and (4). The canonical equations of motion are obtained applying Eq. (7.161) to H:

$$\dot{p}_r = -\frac{\partial H}{\partial r} = -\frac{k}{r^2} + \frac{p_\theta^2}{mr^3} + \frac{p_\phi^2}{mr^3 \sin^2 \theta} \tag{6}$$

$$\dot{p}_\theta = -\frac{\partial H}{\partial \theta} = \frac{p_\phi^2 \cot \theta}{mr^2 \sin^2 \theta} \tag{7}$$

$$\dot{p}_\phi = -\frac{\partial H}{\partial \phi} = 0 \tag{8}$$

The last equation implies that $p_\phi = \text{const}$, which reduces the number of variables on which H depends to four: r, θ, p_r, p_θ:

$$H = \frac{1}{2m}\left[p_r^2 + \frac{p_\theta^2}{r^2} + \frac{\text{const}}{r^2 \sin^2 \theta} \right] - \frac{k}{r} \tag{9}$$

For motion with constant energy, (9) fixes the value of any of the four variables when the other three are given.

From (9), for a given constant value of $H = E$, we obtain

$$\boxed{p_r = \left[2mE - \frac{p_\theta^2 \sin^2 \theta + \text{const}}{r^2 \sin^2 \theta} + \frac{2mk}{r} \right]^{1/2}} \tag{10}$$

and so the projection of the phase path on the $r - p_r$ plane are as shown below.

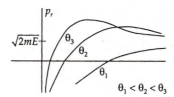

7-35. We use z_i and p_i as our generalized coordinates, the subscript i corresponding to the ith particle. For a uniform field in the z direction the trajectories $z = z(t)$ and momenta $p = p(t)$ are given by

$$\left. \begin{array}{l} z_i = z_{i0} + v_{i0}t - \dfrac{1}{2} gt^2 \\[2mm] p_i = p_{i0} - mgt \end{array} \right] \tag{1}$$

where z_{i0}, p_{i0}, and $v_{i0} = p_{i0}/m$ are the initial displacement, momentum, and velocity of the ith particle.

Using the initial conditions given, we have

$$z_1 = z_0 + \frac{p_0 t}{m} - \frac{1}{2} gt^2 \tag{2a}$$

$$p_1 = p_0 - mgt \tag{2b}$$

$$z_2 = z_0 + \Delta z_0 + \frac{p_0 t}{m} - \frac{1}{2} g t^2 \tag{2c}$$

$$p_2 = p_0 - mgt \tag{2d}$$

$$z_3 = z_0 + \frac{(p_0 + \Delta p_0)t}{m} - \frac{1}{2} g t^2 \tag{2e}$$

$$p_3 = p_0 + \Delta p_0 - mgt \tag{2f}$$

$$z_4 = z_0 + \Delta Z_0 + \frac{(p_0 + \Delta p_0)t}{m} - \frac{1}{2} g t^2 \tag{2g}$$

$$p_4 = p_0 + \Delta p_0 - mgt \tag{2h}$$

The Hamiltonian function corresponding to the ith particle is

$$H_i = T_i + V_i = \frac{m \dot{z}_i^2}{2} + mgz_i = \frac{p_i^2}{2m} + mgz_i = \text{const.} \tag{3}$$

From (3) the phase space diagram for any of the four particles is a parabola as shown below.

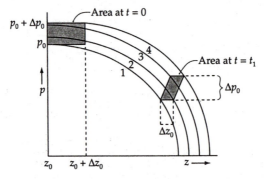

From this diagram (as well as from 2b, 2d, 2f, and 2h) it can be seen that *for any time t,*

$$p_1 = p_2 \tag{4}$$

$$p_3 = p_4 \tag{5}$$

Then for a certain time t the shape of the area described by the representative points will be of the general form

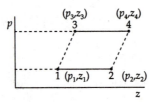

where the base $\overline{1\,2}$ must parallel to the top $\overline{3\,4}$. At time $t = 0$ the area is given by $\Delta z_0 \Delta p_0$, since it corresponds to a rectangle of base Δz_0 and height Δp_0. At any other time the area will be given by

$$A = \left\{ \text{base of parallelogram}\Big|_{t=t_1} = (z_2 - z_1)\Big|_{t=t_1} \right.$$

$$= (z_4 - z_3)\Big|_{t=t_1} = \Delta z_0 \Big\}$$

$$x \left\{ \text{height of parallelogram}\Big|_{t=t_1} = (p_3 - p_1)\Big|_{t=t_1} \right.$$

$$= (p_4 - p_2)\Big|_{t=t_1} = \Delta p_0 \Big\}$$

$$= \Delta p_0 \, \Delta z_0 \qquad (6)$$

Thus, the area occupied in the phase plane is constant in time.

7-41. For small angle of oscillation θ we have

$$T = \frac{1}{2} mb^2 \left(\frac{d\theta}{dt}\right)^2 + \frac{1}{2} m \left(\frac{db}{dt}\right)^2 \quad \text{and} \quad U = -mgb \cos \theta$$

So the Lagrangian reads

$$L = T - U = \frac{1}{2} mb^2 \left(\frac{d\theta}{dt}\right)^2 + \frac{1}{2} m \left(\frac{db}{dt}\right)^2 + mgb \cos \theta$$

from which follow 2 equations of motion

$$\frac{\partial L}{\partial b} = \frac{d}{dt}\frac{\partial L}{\partial \dot{b}} \quad \Rightarrow \quad b\left(\frac{d\theta}{dt}\right)^2 + g \cos \theta = \frac{d^2 b}{dt^2} = -\frac{d\alpha}{dt}$$

$$\frac{\partial L}{\partial \theta} = \frac{d}{dt}\frac{\partial L}{\partial \dot{\theta}} \quad \Rightarrow \quad -mgb \sin \theta = 2mb\frac{db}{dt}\frac{d\theta}{dt} + mb^2\frac{d^2\theta}{dt^2} = -2mb\alpha\frac{d\theta}{dt} + mb^2\frac{d^2\theta}{dt^2}$$

CHAPTER 8

Central-Force Motion

8-3. When $k \to k/2$, the potential energy will decrease to half its former value; but the kinetic energy will remain the same. Since the original orbit is circular, the instantaneous values of T and U are equal to the average values, $\langle T \rangle$ and $\langle U \rangle$. For a $1/r^2$ force, the virial theorem states

$$\langle T \rangle = -\frac{1}{2}\langle U \rangle \tag{1}$$

Hence,

$$E = T + U = -\frac{1}{2}U + U = \frac{1}{2}U \tag{2}$$

Now, consider the energy diagram

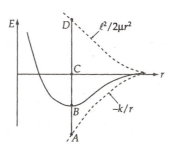

where

$$\overline{CB} = E \qquad \text{original total energy}$$

$$\overline{CA} = U \qquad \text{original potential energy}$$

$$\overline{CD} = U_c \qquad \text{original centrifugal energy}$$

The point B is obtained from $\overline{CB} = \overline{CA} - \overline{CD}$. According to the virial theorem, $E = (1/2)U$ or $\overline{CB} = (1/2)\overline{CA}$. Therefore,

$$\overline{CD} = \overline{CB} = \overline{BA}$$

Hence, if U suddenly is halved, the total energy is raised from B by an amount equal to $(1/2)\overline{CA}$ or by $\overline{CB}$. Thus, the total energy point is raised from B to C; i.e., E(final) = 0 and the orbit is *parabolic*.

8-6.

$$r = |x_2 - x_1|$$

When two particles are initially at rest separated by a distance r_0, the system has the total energy

$$E_0 = -G\frac{m_1 m_2}{r_0} \tag{1}$$

The coordinates of the particles, x_1 and x_2, are measured from the position of the center of mass. At any time the total energy is

$$E = \frac{1}{2}m_1\dot{x}_1^2 + \frac{1}{2}m_2\dot{x}_2^2 - G\frac{m_1 m_2}{r} \tag{2}$$

and the linear momentum, at any time, is

$$p = m_1\dot{x}_1 + m_2\dot{x}_2 = 0 \tag{3}$$

From the conservation of energy we have $E = E_0$, or

$$-G\frac{m_1 m_2}{r_0} = \frac{1}{2}m_1\dot{x}_1^2 + \frac{1}{2}m_2\dot{x}_2^2 - G\frac{m_1 m_2}{r} \tag{4}$$

Using (3) in (4), we find

$$\boxed{\begin{aligned} \dot{x}_1 = v_1 = m_2\sqrt{\frac{2G}{M}\left[\frac{1}{r} - \frac{1}{r_0}\right]} \\[2em] \dot{x}_1 = v_2 = -m_1\sqrt{\frac{2G}{M}\left[\frac{1}{r} - \frac{1}{r_0}\right]} \end{aligned}} \tag{5}$$

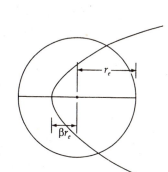

The orbit of the comet is a parabola ($\varepsilon = 1$), so that the equation of the orbit is

$$\frac{\alpha}{r} = 1 + \cos\theta \tag{1}$$

We choose to measure θ from perihelion; hence

$$r(\theta = 0) = \beta r_E \tag{2}$$

Therefore,

$$\alpha = \frac{\ell^2}{\mu k} = 2\beta r_E \tag{3}$$

Since the total energy is zero (the orbit is parabolic) and the potential energy is $U = -k/r$, the time spent within the orbit of the Earth is

$$T = 2\int_{\beta r_E}^{r_E} \frac{dr}{\sqrt{\frac{2}{\mu}\left[\frac{k}{r} - \frac{\ell^2}{2\mu r^2}\right]}}$$

$$= \sqrt{\frac{2\mu}{k}} \int_{\beta r_E}^{r_E} \frac{r\,dr}{\sqrt{r - \beta r_E}} \tag{4}$$

$$= \sqrt{\frac{2\mu}{k}}\left[-\frac{2(-2\beta r_E - r)}{3}\sqrt{r - \beta r_E}\right]_{\beta r_E}^{r_E}$$

from which

$$T = \sqrt{\frac{2\mu}{k}}\left[\frac{2}{3} r_E^{3/2}\,(2\beta + 1)\sqrt{1 - \beta}\right] \tag{5}$$

Now, the period and the radius of the Earth are related by

$$\tau_E^2 = \frac{4\pi^2 \mu_E}{k'} r_E^3 \tag{6}$$

or,

$$r_E^{3/2} = \sqrt{\frac{k'}{\mu_E}} \frac{\tau_E}{2\pi} \tag{7}$$

Substituting (7) into (5), we find

$$T = \sqrt{\frac{2\mu}{k}} \frac{2}{3} \sqrt{\frac{k'}{\mu_E} \frac{\tau_E}{2\pi}} (2\beta+1) \sqrt{1-\beta} \tag{8}$$

where $k = GM_s\mu$ and $k' = GM_s\mu_E$. Therefore,

$$\boxed{T = \frac{1}{3\pi} \sqrt{2(1-\beta)}\,(1+2\beta)\,\tau_E} \tag{9}$$

where $\tau_E = 1$ year. Now, $\beta = r_{\text{Mercury}}/r_E = 0.387$. Therefore,

$$T = \frac{1}{3\pi} \sqrt{2(1-0.387)}\,(1+2\times0.387)\times 365 \text{ days}$$

so that

$$\boxed{T = 76 \text{ days}} \tag{10}$$

8-16.   The total energy of the particle is

$$E = T + U \tag{1}$$

a principle that by no means pushes the philosophical envelope of physical interpretation. The impulse that causes $v \to v + \delta v$ changes the kinetic energy, not the potential energy. We therefore have

$$\delta E = \delta T = \delta\left(\frac{1}{2}mv^2\right) = mv\,\delta v \tag{2}$$

By the virial theorem, for a nearly circular orbit we have

$$E = -\frac{1}{2}mv^2 \tag{3}$$

so that

$$\frac{\delta E}{-E} = \frac{2\delta v}{v} \tag{4}$$

where we have written $-E$ since $E < 0$. The major and minor axes of the orbit are given by

$$a = -\frac{k}{2E} \qquad b = \frac{\ell}{\sqrt{-2\mu E}} \tag{5}$$

Now let us compute the changes in these quantities. For a we have

$$\delta a = -\delta\left(\frac{k}{2E}\right) = \frac{k\delta E}{2E^2} = a\left(\frac{\delta E}{-E}\right) \tag{6}$$

and for b we have

$$\delta b = \delta\left[\frac{\ell}{\sqrt{-2\mu E}}\right] = \frac{\delta\ell}{\sqrt{-2\mu E}} + \frac{\ell}{\sqrt{-2\mu E^3}}\left[-\frac{1}{2}\delta E\right] = b\left[\frac{\delta\ell}{\ell} - \frac{\delta E}{2E}\right] \tag{7}$$

Easily enough, we can show that $\delta\ell/\ell = \delta v/v$ and therefore

$$\frac{\delta a}{a} = \frac{\delta b}{b} = \frac{\delta E}{-E} = \frac{2\delta v}{v} \tag{8}$$

8-23.   Start with the equation of the orbit:

$$\frac{\alpha}{r} = 1 + \varepsilon\cos\theta \tag{1}$$

and take its time derivative

$$\frac{\dot{r}}{r^2} = \frac{\varepsilon}{\alpha}\dot{\theta}\sin\theta = \frac{\varepsilon\ell}{\alpha\mu r^2}\sin\theta \tag{2}$$

Now from Equation (8.45) and (8.43) we have

$$\tau = \frac{2\mu}{\ell}\cdot\pi ab = \frac{2\pi\mu a\alpha}{\ell\sqrt{1-\varepsilon^2}} \tag{3}$$

so that from (2)

$$|\dot{r}|_{max} = \frac{\varepsilon}{\mu}\cdot\frac{\ell}{\alpha} = \frac{2\pi a\varepsilon}{\tau\sqrt{1-\varepsilon^2}} \tag{4}$$

as desired.

8-26.

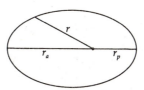

First, consider a velocity kick Δv applied along the direction of travel at an arbitrary place in the orbit. We seek the optimum location to apply the kick.

$$E_1 = \text{initial energy}$$

$$= \frac{1}{2}mv^2 - \frac{GMm}{r}$$

$$E_2 = \text{final energy}$$

$$= \frac{1}{2}m(v+\Delta v)^2 - \frac{GMm}{r}$$

We seek to maximize the energy gain $E_2 - E_1$:

$$E_2 - E_1 = \frac{1}{2}m\left(2v\,\Delta v + \Delta v^2\right)$$

For a given Δv, this quantity is clearly a maximum when v is a maximum; i.e., at perigee.

Now consider a velocity kick ΔV applied at perigee in an arbitrary direction:

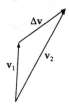

The final energy is

$$\frac{1}{2}mv_2^2 = \frac{GMm}{r_p}$$

This will be a maximum for a maximum $|v_2|$; which clearly occurs when v_1 and Δv are along the same direction.

> Thus, the most efficient way to change the energy of an elliptical orbit (for a single engine thrust) is by firing along the direction of travel at perigee.

8-31.   From the given force, we find

$$\frac{dF(r)}{dr} = F'(r) = \frac{2k}{r^3} + \frac{4k'}{r^5} \tag{1}$$

Therefore, the condition of stability becomes [see Eq. (8.93)]

$$\frac{F'(\rho)}{F(\rho)} + \frac{3}{\rho} = \frac{\dfrac{2}{\rho^5}\left(k\rho^2 + 2k'\right)}{-\dfrac{1}{\rho^4}\left(k\rho^2 + k'\right)} + \frac{3}{\rho} > 0 \tag{2}$$

or,

$$\frac{k\rho^2 - k'}{\rho\left(k\rho^2 + k'\right)} > 0 \tag{3}$$

Therefore, if $\rho^2 k > k'$, the orbit is stable.

8-34.   The total energy of the system is

$$E = \frac{1}{2}m\left(\dot{r}^2 + r^2\dot{\theta}^2 + \dot{r}^2\cot^2\alpha\right) + mgr\cot\alpha \tag{1}$$

or,

$$E = \frac{1}{2}m\left(1 + \cot^2 \alpha\right)\dot{r}^2 + \frac{1}{2}mr^2\dot{\theta}^2 + mgr\cot\alpha \qquad (2)$$

Substituting $\ell = mr^2\dot{\theta}$, we have

$$E = \frac{1}{2}m\left(1 + \cot^2 \alpha\right)\dot{r}^2 + \frac{\ell^2}{2mr^2} + mgr\cot\alpha \qquad (3)$$

Therefore, the effective potential is

$$V(r) = \frac{\ell^2}{2mr^2} + mgr\cot\alpha \qquad (4)$$

At the turning point we have $\dot{r} = 0$, and (3) becomes a cubic equation in r:

$$mgr^3\cot\alpha - Er^2 + \frac{\ell^2}{2m} = 0 \qquad (5)$$

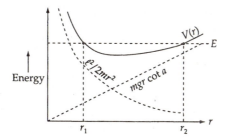

This cubic equation has three roots. If we attempt to find these roots graphically from the intersections of $E = $ const. and $V(r) = \ell^2/2mr^2 + mgr\cot\alpha$, we discover that only two of the roots are real. (The third root is imaginary.) These two roots specify the planes between which the motion takes place.

8-39. We must calculate the quantity Δv_1 for transfers to Venus and Mars. From Eqs. (8.54), (8.53), and (8.51):

$$\Delta v_1 = v_{t_1} - v_1$$

$$= \sqrt{\frac{2k}{mr_1}\left[\frac{r_2}{r_1 + r_2}\right]} - \sqrt{\frac{k}{mr_1}}$$

where

$$\frac{k}{m} = GM_s = \left(6.67 \times 10^{-11} \text{ m}^3/\text{s}^2\text{-kg}\right)\left(1.99 \times 10^{30} \text{ kg}\right)$$

$$r_1 = \text{mean Earth-sun distance} = 150 \times 10^9 \text{ m}$$

$$r_2 = \text{mean} \begin{bmatrix} \text{Venus} \\ \text{Mars} \end{bmatrix} - \text{sun distance} = \begin{bmatrix} 108 \\ 228 \end{bmatrix} \times 10^9 \text{ m}$$

Substituting gives

$$\Delta v_{\text{Venus}} = -2.53 \text{ km/sec}$$

$$\Delta v_{\text{Mars}} = 2.92 \text{ km/sec}$$

where the negative sign for Venus means the velocity kick is opposite to the Earth's orbital motion.

> Thus, a Mars flyby requires a larger Δv than a Venus flyby.

8-41.   From the equations in Section 8.8 regarding Hohmann transfers

$$\Delta v = \Delta v_1 + \Delta v_2$$

$$= v_{t_1} - v_1 + v_2 - v_{t_2}$$

where

$$v_{t_1} = \sqrt{\frac{2k}{mr_1}\left[\frac{r_2}{r_1 + r_2}\right]}; \quad v_1 = \sqrt{\frac{k}{mr_1}}$$

$$v_{t_2} = \sqrt{\frac{2k}{mr_2}\left[\frac{r_1}{r_1 + r_2}\right]}; \quad v_1 = \sqrt{\frac{k}{mr_2}}$$

Substituting

$$\frac{k}{m} = GM_e = \left(6.67 \times 10^{-11} \text{ Nm}^2/\text{kg}^2\right)\left(5.98 \times 10^{24} \text{ kg}\right)$$

$$r_1 = 200 \text{ km} + r_e = 6.37 \times 10^6 \text{ m} + 2 \times 10^5 \text{ m}$$

$$r_2 = \text{mean Earth-moon distance} = 3.84 \times 10^8 \text{ m}$$

gives

> $$\Delta v = 3966 \text{ m/s}$$

From Eq. (8.58), the time of transfer is given by

$$T = \tau \sqrt{\frac{m}{k}} \, a_t^{3/2} = \pi \sqrt{\frac{m}{k}} \left[\frac{r_1 + r_2}{2}\right]^{3/2}$$

Substituting gives

> $$\tau = 429,000 \text{ sec.} \approx 5 \text{ days}$$

8-46.   In equilibrium circular orbit,

$$\frac{Mv^2}{R} = \frac{GM^2}{4R^2} \quad \Rightarrow \quad R = \frac{GM}{4v^2}$$

where M is the Sun's mass.

The period is

$$T = \frac{2\pi R}{v} = \frac{4\pi R\sqrt{R}}{\sqrt{GM}} = \frac{\sqrt{2}\pi D^{3/2}}{\sqrt{GM}} \approx 9 \times 10^7 \, \text{yr}$$

where $D = 2R$ is the separation distance of 2 stars.

Dynamics of a System of Particles

9-3.

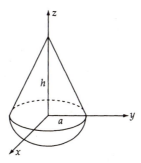

By symmetry, $\bar{x} = \bar{y} = 0$.

From problem 9-2, the center of mass of the cone is at $z = \dfrac{1}{4}h$.

From problem 9-1, the center of mass of the hemisphere is at

$$z = -\frac{3}{8}a\left(r_2 = a, r_1 = 0\right)$$

So the problem reduces to

- $z_1 = \dfrac{1}{4}h;\; m_1 = \rho_1 \dfrac{1}{3}\pi a^2 h$

- $z_2 = -\dfrac{3}{8}a;\; m_2 = \rho_2 \dfrac{2}{3}\pi a^3$

$$\bar{z} = \frac{m_1 z_1 + m_2 z_2}{m_1 + m_2} = \frac{\rho_1 h^2 - 3\rho_2 a^2}{4\left(\rho_1 h + 2\rho_2 a\right)}$$

for $\rho_1 = \rho_2$

$$\boxed{\bar{z} = \frac{h^2 - 3a^3}{4(2a + h)}}$$

9-8. By symmetry, $\boxed{\bar{x} = 0}$. Also, by symmetry, we may integrate over the $x > 0$ half of the triangle to get $\bar{y}$. $\sigma = $ mass/area

$$\bar{y} = \frac{\int_{x=0}^{\frac{a}{\sqrt{2}}} \int_{y=0}^{\frac{a}{\sqrt{2}}-x} \sigma y \, dy \, dx}{\int_{x=0}^{\frac{a}{\sqrt{2}}} \int_{y=0}^{\frac{a}{\sqrt{2}}-x} \sigma \, dy \, dx} = \frac{a}{3\sqrt{2}}$$

$$\boxed{\bar{y} = \frac{a}{3\sqrt{2}}}$$

9-15. $\qquad\qquad\qquad \sigma = $ mass/length

$$F = \frac{dp}{dt} \text{ becomes}$$

$$mg = m\dot{v} + \dot{m}v$$

where m is the mass of length x of the rope. So

$$m = \sigma x; \, \dot{m} = \sigma \dot{x}$$

$$\sigma x g = \sigma x \frac{dv}{dt} + \sigma \dot{x} v$$

$$x g = x \frac{dv}{dx}\frac{dx}{dt} + v^2$$

$$x g = xv \frac{dv}{dx} + v^2$$

Try a power law solution:

$$v = ax^n; \frac{dv}{dx} = nax^{n-1}$$

Substituting,

$$x g = x\left(ax^n\right)\left(nax^{n-1}\right) + a^2 x^{2n}$$

or

$$x g = a^2 (n+1)x^{2n}$$

Since this must be true for all x, the exponent and coefficient of x must be the same on both sides of the equation.

Thus we have: $1 = 2n$ or $n = \dfrac{1}{2}$

$$g = a^2 (n+1) \text{ or } a = \sqrt{\frac{2g}{3}}$$

So

$$\boxed{v = \sqrt{\frac{2gx}{3}}}$$

$$a = \frac{dv}{dt} = \frac{dv}{dx}\frac{dx}{dt} = v\frac{dv}{dx} = \left[\frac{2gx}{3}\right]^{1/2} \frac{g}{3}\left[\frac{2gx}{3}\right]^{-1/2}$$

$$\boxed{a = \frac{g}{3}}$$

$$T_i = 0 \qquad U_i = 0 \qquad (y = 0 \text{ on table})$$

$$T_f = \frac{1}{2}mv^2 = \frac{1}{2}m\left[\frac{2gL}{3}\right] = \frac{mgL}{3}$$

$$U_f = mgh = -mg\frac{L}{2}$$

So $E_i = 0;\ E_f = -\dfrac{mgL}{6}$

$$\boxed{\text{Energy lost} = \frac{mgL}{6}}$$

9-18.   Once we have solved Problem 9-17, it becomes an easy matter to write the expression for the tension (Equation 9.18):

$$\frac{T}{mg} = \frac{1 + 2\alpha - 6\alpha^2}{2(1 - 2\alpha)} \tag{1}$$

This is plotted vs. the natural time using the solution of Problem 9-17.

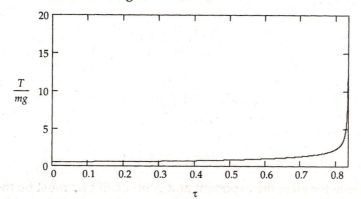

9-20.

Let ρ = mass/length

The force on the rope is due to gravity

$$F = (a+x)\rho g - (a-x)\rho g$$

$$= 2x\rho g$$

$$\frac{dp}{dt} = m\frac{dv}{dt} = 2a\rho\frac{dv}{dt}$$

So $F = \dfrac{dp}{dt}$ becomes

$$xg = a\frac{dv}{dt}$$

Now

$$\frac{dv}{dt} = \frac{dv}{dx}\cdot\frac{dx}{dt} = v\frac{dv}{dx}$$

So

$$xg = av\frac{dv}{dx}$$

or

$$vdv = \frac{g}{a}xdx$$

Integrating yields

$$\frac{1}{2}v^2 = \frac{g}{2a}x^2 + c$$

Since $v = 0$ when $x = 0$, $c = 0$.

Thus

$$v^2 = \frac{g}{a}x^2$$

When the rope clears the nail, $x = a$. Thus

$$\boxed{v = \sqrt{ga}}$$

9-24.

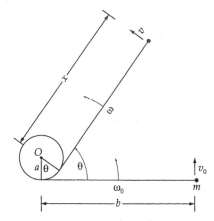

The energy of the system is, of course, conserved, and so we have the following relation involving the instantaneous velocity of the particle:

$$\frac{1}{2}mv^2 = \frac{1}{2}mv_0^2 \tag{1}$$

The angular momentum about the center of the cylinder is not conserved since the tension in the string causes a torque. Note that although the velocity of the particle has both radial and angular components, there is only one independent variable, which we chose to be θ. Here $\omega = \dot{\theta}$ is the angular velocity of the particle about the point of contact, which also happens to be the rate at which the point of contact is rotating about the center of the cylinder. Hence we may write

$$v_0 = \omega_0 b; \qquad v = \omega_0 (b - a\theta) \tag{2}$$

From (1) and (2), we can solve for the angular velocity after turning through an angle θ

$$\omega = \frac{\omega_0}{1 - \dfrac{a}{b}\theta} \tag{3}$$

The tension will then be (look at the point of contact)

$$T = m\omega^2 (b - a\theta) = m\omega_0\omega b \tag{4}$$

9-28. Using the notation from the chapter:

$$m_1: \quad T_i = T_0, \quad T_f = T_1$$

$$m_2: \quad T_i = 0; \quad T_f = T_2$$

Thus

$$T_0 = T_1 + T_2 \quad or \quad 1 = \frac{T_1}{T_0} + \frac{T_2}{T_0} \tag{1}$$

If we want the kinetic energy loss for m_1 to be a maximum, we must minimize $\dfrac{T_1}{T_0}$ or, equivalently, maximize $\dfrac{T_2}{T_0}$.

From Eq. (9.88):

$$\frac{T_2}{T_0} = \frac{4m_1 m_2}{\left(m_1 + m_2\right)^2} \cos^2 \zeta$$

To maximize this, $\zeta = 0$ (it can't $= 180°$).

$$\frac{T_2}{T_0} = \frac{4m_1 m_2}{\left(m_1 + m_2\right)^2}$$

The kinetic energy loss for m_1 is $T_0 - T_1$. The fraction of kinetic energy loss is thus

$$\frac{T_0 - T_1}{T_0} = 1 - \frac{T_1}{T_0} = \frac{T_2}{T_0} \text{ (from (1))}$$

$$\boxed{\left.\frac{T_0 - T_1}{T_0}\right|_{\text{max}} = \frac{4m_1 m_2}{\left(m_1 + m_2\right)^2}}$$

$\zeta = 0$ implies $\psi = 0, 180°$ (conservation of p_v). So the reaction is as follows

Before: $\underset{m_1}{O} \xrightarrow{v_1} \underset{m_2}{O}$

After: $\underset{m_1}{O} \xrightarrow{v_1} \quad \underset{m_2}{O} \xrightarrow{v_2}$

$p_x: \quad m_1 v = m_1 v_1 + m_2 v_2$

$E: \quad \dfrac{1}{2} m_1 v^2 = \dfrac{1}{2} m_1 v_1^2 + \dfrac{1}{2} m_2 v_2^2$

Solving for v_1 gives $v_1 = \dfrac{m_1 - m_2}{m_1 + m_2} v$

So

m_2 travels in $+ x$ direction

m_1 travels in $\left[\begin{array}{l} + x \text{ direction if } m_1 > m_2 \\[1mm] - x \text{ direction if } m_1 < m_2 \end{array}\right.$

9-34.

Cons. of p_z: $mu_1 = mv_1 \cos 45° + mv_2 \cos \theta$ (1)

Cons. of p_y: $0 = mv_1 \sin 45° - mv_2 \sin \theta$ (2)

Cons. of energy (elastic collision)

$$\frac{1}{2} mu_1^2 = \frac{1}{2} mv_1^2 - \frac{1}{2} mv_2^2$$ (3)

Solve (1) for $\cos \theta$:

$$\cos \theta = \frac{u_1 - v_1/\sqrt{2}}{v_2}$$

Solve (2) for $\sin \theta$:

$$\sin \theta = \frac{v_1}{\sqrt{2}\, v_2}$$

Substitute into $\cos^2 \theta + \sin^2 \theta = 1$, simplify, and the result is

$$u_1^2 = v_2^2 - v_1^2 + \sqrt{2}\, u_1 v_1$$

Combining this with (3) gives

$$2v_1^2 = \sqrt{2}\, u_1 v_1$$

We are told $v_1 \neq 0$, hence

$$\boxed{v_1 = u_1/\sqrt{2}}$$

Substitute into (3) and the result is

$$\boxed{v_2 = u_1/\sqrt{2}}$$

Since $v_1 = v_2$, (2) implies

$$\boxed{\theta = 45°}$$

9-35.   From the following two expressions for T_1/T_0 ,

$$\frac{T_1}{T_0} = \frac{v_1^2}{u_1^2} \qquad \text{Eq. (9.82)}$$

$$\frac{T_1}{T} = \frac{m_1^2}{(m_1 + m_2)^2} \left[\cos \psi \pm \sqrt{\left[\frac{m_2}{m_1}\right]^2 - \sin^2 \psi} \right]^2 \qquad \text{Eq. (9.87b)}$$

we can find the expression for the final velocity v_1 of m_1 in the lab system in terms of the scattering angle ψ:

$$v_1 = \frac{m_1 u_1}{m_1 + m_2} \left[\cos \psi \pm \sqrt{\left[\frac{m_2}{m_1}\right]^2 - \sin^2 \psi} \right] \qquad (1)$$

If time is to be constant on a certain surface that is a distance r from the point of collision, we have

$$r = v_1 t_0 \qquad (2)$$

Thus,

$$r = \frac{m_1 u_1 t_0}{m_1 + m_2} \left[\cos \psi \pm \sqrt{\left[\frac{m_2}{m_1}\right]^2 - \sin^2 \psi} \right] \qquad (3)$$

This is the equation of the required surface. Let us consider the following cases:

i) $m_2 = m_1$:

$$r = \frac{u_1 t_0}{2} \left[\cos \psi \pm \sqrt{1 - \sin^2 \psi} \right] = u_1 t_0 \cos \psi \qquad (4)$$

(The possibility $r = 0$ is uninteresting.)

ii) $m_2 = 2m_1$:

$$r = \frac{u_1 t_0}{3} \left[\cos \psi \pm \sqrt{4 - \sin^2 \psi} \right] \qquad (5)$$

iii) $m_2 = \infty$: Rewriting (3) as

$$r = \frac{m_1 u_1 t_0}{1 + \dfrac{m_1}{m_2}} \left[\frac{\cos \psi}{m_2} \pm \sqrt{\left[\frac{1}{m_1}\right]^2 - \frac{\sin^2 \psi}{m_2^2}} \right] \qquad (6)$$

and taking the limit $m_2 \to \infty$, we find

$$r = u_1 t_0 \qquad (7)$$

All three cases yield spherical surfaces, but with the centers displaced:

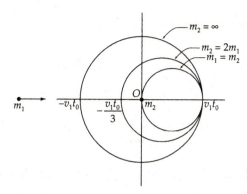

This result is useful in the design of a certain type of nuclear detector. If a hydrogenous material is placed at 0 then for neutrons incident on the material, we have the case $m_1 = m_2$. Therefore, neutrons scattered from the hydrogenous target will arrive on the surface A with the same time delay between scattering and arrival, independent of the scattering angle. Therefore, a coincidence experiment in which the time delay is measured can determine the energies of the incident neutrons. Since the entire surface A can be used, a very efficient detector can be constructed.

9-41. Using $y = v_0 t - \dfrac{1}{2} g t^2$ and $v = v_0 - gt$, we can get the velocities before and after the collision:

Before: $u_1 = -gt_1$ where $h_1 = \dfrac{1}{2} g t_1^2$

So $u_1 = -g\sqrt{\dfrac{2h_1}{g}} = -\sqrt{2gh_1}$

After: $0 = v_0 - gt_2$ or $t_2 = v_0/g$

$h_2 = v_0 t_2 - \dfrac{1}{2} g t_2^2$

$= \dfrac{v_0^2}{g} - \dfrac{1}{2}\dfrac{v_0^2}{g}$ or $v_0 = \sqrt{2gh_2}$

So $v_1 = \sqrt{2gh_2}$

Thus

$$\varepsilon = \frac{|v_2 - v_1|}{|u_2 - u_1|} = \frac{\sqrt{2gh_2}}{\sqrt{2gh_1}}$$

$$\boxed{\varepsilon = \sqrt{\frac{h_2}{h_1}}}$$

$$T_{\text{lost}} = T_i - T_f$$

$$\text{Fraction lost} = \frac{T_i - T_f}{T_i}$$

$$= \frac{u_1^2 - v_1^2}{u_1^2} = \frac{h_1 - h_2}{h_1} = 1 - \frac{h_2}{h_1}$$

$$\boxed{\frac{T_i - T_f}{T_i} = 1 - \varepsilon^2}$$

9-45.   Since the total number of particles scattered into a unit solid angle must be the same in the lab system as in the CM system [cf. Eq. (9.124) in the text],

$$\sigma(\theta)\, 2\pi \sin\theta\, d\theta = \sigma(\psi) \cdot 2\pi \sin\psi\, d\psi \tag{1}$$

Thus,

$$\sigma(\theta) = \sigma(\psi) \frac{\sin\psi}{\sin\theta} \frac{d\psi}{d\theta} \tag{2}$$

The relation between θ and ψ is given by Eq. (9.69), which is

$$\tan\psi = \frac{\sin\theta}{\cos\theta + x} \tag{3}$$

where $x = m_1/m_2$. Using this relation, we can eliminate ψ from (2):

$$\sin\psi = \frac{1}{\sqrt{1 + \dfrac{1}{\tan^2\psi}}} = \frac{1}{\sqrt{1 + \dfrac{(\cos\theta + x)^2}{\sin^2\theta}}} = \frac{\sin\theta}{\sqrt{1 + 2x\cos\theta + x^2}} \tag{4}$$

$$\frac{d\psi}{d\theta} = \frac{d\psi}{d(\tan\psi)} \frac{d(\tan\psi)}{d\theta} = \cos^2\psi \frac{\cos\theta(\cos\theta + x) + \sin^2\theta}{(\cos\theta + x)^2} \tag{5}$$

Since $\cos^2\psi = \dfrac{1}{1 + \tan^2\psi}$, (5) becomes

$$\frac{d\psi}{d\theta} = \frac{1}{1 + \dfrac{\sin^2\theta}{(\cos\theta + x)^2}} \frac{1 + x\cos\theta}{(\cos\theta + x)^2} = \frac{1 + x\cos\theta}{1 + 2x\cos\theta + x^2} \tag{6}$$

Substituting (4) and (6) into (2), we find

$$\boxed{\sigma(\theta) = \sigma(\psi) \frac{1 + x\cos\theta}{\left(1 + 2x\cos\theta + x^2\right)^{3/2}}} \tag{7}$$

9-49. The differential cross section for Rutherford scattering in the CM system is [cf. Eq. (9.140) in the text]

$$\sigma(\theta) = \frac{k^2}{16T_0'^2} \frac{1}{\sin^4 \frac{\theta}{2}} \tag{1}$$

where [cf. Eq. (9.79)]

$$T_0' = \frac{m_2}{m_1 + m_2} T_0 \tag{2}$$

Thus,

$$\sigma(\theta) = \frac{k^2}{16T_0^2} \frac{1}{\sin^4 \frac{\theta}{2}} \left[\frac{m_1 + m_2}{m_2} \right]^2$$

$$= \frac{k^2}{16T_0^2} \frac{1}{\sin^4 \frac{\theta}{2}} \left[1 + \frac{m_1}{m_2} \right]^2 \tag{3}$$

Since $m_1/m_2 \ll 1$, we expand

$$\left[1 + \frac{m_1}{m_2} \right]^2 \cong 1 + 2\frac{m_1}{m_2} + \dots \tag{4}$$

Thus, to the first order in m_1/m_2, we have

$$\boxed{\sigma(\theta) = \frac{k^2}{16T_0^2} \frac{1}{\sin^4 \frac{\theta}{2}} \left[1 + 2\frac{m_1}{m_2} \right]} \tag{5}$$

This result is the same as Eq. (9.140) except for the correction term proportional to m_1/m_2.

9-55. The velocity equation (9.165) gives us:

$$v(t) = -gt + u \ln\left[\frac{m_0}{m(t)} \right] \tag{1}$$

where $m(t) = m_0 - \alpha t$, the burn rate $\alpha = 9m_0/10\tau$, the burn time $\tau = 300 \text{ s}$, and the exhaust velocity $u = 4500 \text{ m} \cdot \text{s}^{-1}$. These equations are good only from $t = 0$ to $t = \tau$. First, let us check that the rocket does indeed lift off at $t = 0$: the thrust $\alpha u = 9um_0/10\tau = 13.5 \text{ m} \cdot \text{s}^{-2} \cdot m_0 > m_0 g$, as required. To find the maximum velocity of the rocket, we need to check it at the times $t = 0$ and $t = \tau$, and also check for the presence of any extrema in the region $0 < t < \tau$. We have $v(0) = 0$, $v(\tau) = -g\tau + u \ln 10 = 7400 \text{ m} \cdot \text{s}^{-1}$, and calculate

$$\frac{dv}{dt} = -g + \frac{\alpha u}{m(t)} = g\left[\frac{\alpha u}{m(t)g} - 1\right] > 0 \tag{2}$$

The inequality follows since $\alpha u > m_0 g > m(t)g$. Therefore the maximum velocity occurs at $t = \tau$, where $v = -g\tau + u \ln 10 = 7400 \text{ m} \cdot \text{s}^{-1}$. A similar single-stage rocket cannot reach the moon since $v(t) < u \ln\left(m_0/m_{final}\right) = u \ln 10 \approx 10.4 \text{ m} \cdot \text{s}^{-1}$, which is less than escape velocity and independent of fuel burn rate.

9-56.

a) Since the rate of change of mass of the droplet is proportional to its cross-sectional area, we have

$$\frac{dm}{dt} = k\pi r^2 \tag{1}$$

If the density of the droplet is ρ,

$$m = \frac{4\pi}{3}\rho r^3 \tag{2}$$

so that

$$\frac{dm}{dt} = \frac{dm}{dr}\frac{dr}{dt} = 4\pi\rho r^2 \frac{dr}{dt} = \pi k r^2 \tag{3}$$

Therefore,

$$\frac{dr}{dt} = \frac{k}{4\rho} \tag{4}$$

or,

$$\boxed{r = r_0 + \frac{k}{4\rho}t} \tag{5}$$

as required.

b) The mass changes with time, so the equation of motion is

$$F = \frac{d}{dt}(mv) = m\frac{dv}{dt} + v\frac{dm}{dt} = mg \tag{6}$$

Using (1) and (2) this becomes

$$\frac{4\pi}{3}\rho r^3 \frac{dv}{dt} + \pi k r^2 v = \frac{4\pi}{3}\rho r^3 g \tag{7}$$

or,

$$\frac{dv}{dt} + \frac{3k}{4\rho r}v = g \tag{8}$$

Using (5) this becomes

$$\frac{dv}{dt} + \frac{3k}{4\rho} \frac{v}{r_0 + \frac{k}{4\rho}t} = g \tag{9}$$

If we set $A = \dfrac{3k}{4\rho}$ and $B = \dfrac{k}{4\rho}$, this equation becomes

$$\frac{dv}{dt} + \frac{A}{r_0 + Bt} v = g \tag{10}$$

and we recognize a standard form for a first-order differential equation:

$$\frac{dv}{dt} + P(t)v = Q(t) \tag{11}$$

in which we identify

$$P(t) = \frac{A}{r_0 + Bt}; \quad Q(t) = g \tag{12}$$

The solution of (11) is

$$v(t) = e^{-\int P(t)\,dt} \left[\int e^{\int P(t)\,dt} Q\,dt + \text{constant} \right] \tag{13}$$

Now,

$$\int P(t)\,dt = \int \frac{A}{r_0 + Bt}\,dt = \frac{A}{B}\ln(r_0 + Bt)$$

$$= \ln(r_0 + Bt)^3 \tag{14}$$

since $\dfrac{A}{B} = 3$. Therefore,

$$e^{\int P\,dt} = (r_0 + Bt)^3 \tag{15}$$

Thus,

$$v(t) = (r_0 + Bt)^{-3} \left[\int (r_0 + Bt)^3 g\,dt + \text{constant} \right]$$

$$= (r_0 + Bt)^{-3} \left[\frac{g}{4B}(r_0 + Bt)^4 + C \right] \tag{16}$$

The constant C can be evaluated by setting $v(t = 0) = v_0$:

$$v_0 = \frac{1}{r_0^3} \left[\frac{g}{4B}(r_0^4 + C) \right] \tag{17}$$

so that

$$C = v_0 r_0^3 - \frac{g}{4B} r_0^4 \tag{18}$$

We then have

$$v(t) = \frac{1}{(r_0 + Bt)^3} \left[\frac{g}{4B}(r_0 + Bt)^4 + v_0 r_0^3 - \frac{g}{4B} r_0^4 \right] \tag{19}$$

or,

$$v(t) = \frac{1}{(Bt)^3} \left[\frac{g}{4B}(Bt)^4 + 0(r_0^3) \right] \tag{20}$$

where $0(r_0^3)$ means "terms of order r_0^3 and higher." If r_0 is sufficiently small so that we can neglect these terms, we have

$$\boxed{v(t) \propto t} \tag{21}$$

as required.

9-62. To hover above the surface requires the thrust to counteract the gravitational force of the moon. Thus:

$$-u \frac{dm}{dt} = \frac{1}{6} mg$$

$$-\frac{6u}{g} \frac{dm}{m} = dt$$

Integrate from $m = m_0$ to $0.8\, m_0$ and $t = 0$ to T:

$$T = -\frac{6u}{g} \ln 0.8 = -\frac{6(2000 \text{ m/s})}{9.8 \text{ m/s}^2} \ln 0.8$$

$$\boxed{T = 273 \text{ sec}}$$

9-64. We start with the equation of motion for a rocket influenced by an external force, Eq. (9.160), with F_{ext} including gravity, and later, air resistance.

a) There is only constant acceleration due to gravity to worry about, so the problem can be solved analytically. From Eq. (9.166), we can obtain the rocket's height at burnout

$$y_b = ut_b - \frac{1}{2} gt_b^2 - \frac{mu}{\alpha} \ln \left[\frac{m_0}{m_b} \right] \tag{1}$$

where m_b is the mass of the rocket at burnout and $\alpha = (m_0 - m_b)/t_b$. Substitution of the given values gives $y_b \approx 250 \text{ km}$. After burnout, the rocket travels an additional $v_b^2 / 2g$, where v_b is the rocket velocity at burnout. The final height the rocket ends up being $\approx 3700 \text{ km}$, after everything is taken into account.

b) The situation, and hence the differential equation, becomes more complicated when air resistance is added. Substituting $F_{ext} = -mg - c_W \rho A v^2 / 2$ (with $\rho = 1.3 \text{ kg} \cdot \text{m}^{-3}$) into Equation (9.160), we obtain

$$\frac{dv}{dt} = \frac{u\alpha}{m} - g - \frac{c_W \rho A v^2}{2m} \tag{2}$$

We must remember that the mass m is also a function of time, and we must therefore include it also in the system of equations. To be specific, the system of equations we must use to do this by computer are

$$\begin{bmatrix} \dot{y} \\ \dot{v} \\ \dot{m} \end{bmatrix} = \begin{bmatrix} v \\ \dfrac{u\alpha}{m} - g - \dfrac{c_W \rho A v^2}{2m} \\ -\alpha \end{bmatrix} \tag{3}$$

These must be integrated from the beginning until the burnout time, and therefore must be integrated with the substitution $\alpha = 0$. Firstly, we get the velocity and height at burnout to be $v_b \simeq 7000 \text{ m} \cdot \text{s}^{-1}$ and $y_b \simeq 230 \text{ km}$. We can numerically integrate to get the second part of the journey, or use the results of Problem 9-63(b) to help us get the additional distance travelled with air resistance, analytically. The total height to which the rocket rises is $\simeq 890 \text{ km}$ in a total flight time of $\simeq 410 \text{ s}$.

c) The variation in the acceleration of gravity is taken into account by substituting $GM_e / (R_e + y)^2 = g \left[R_e / (R_e + y) \right]^2$ for g in the differential equation in part (b). This gives $v_b \simeq 6900 \text{ m} \cdot \text{s}^{-1}$, $y_b \simeq 230 \text{ km}$, with total height $\simeq 950 \text{ km}$ and time-of-flight $\simeq 460 \text{ s}$.

d) Now one simply substitutes the given expression for the air density, $\rho(y)$ for ρ, into the differential equation from part (c). This gives $v_b \simeq 8200 \text{ m} \cdot \text{s}^{-1}$, $y_b \simeq 250 \text{ km}$, and total height $\simeq 8900 \text{ km}$ with time-of-flight $\simeq 2900 \text{ s}$.

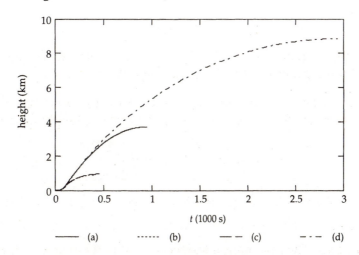

CHAPTER 10

Motion in a Noninertial Reference Frame

10-2. The fixed frame is the ground.

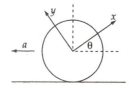

The rotating frame has the origin at the center of the tire and is the frame in which the tire is at rest.

From Eqs. (10.24), (10.25):

$$\mathbf{a}_f = \ddot{\mathbf{R}}_f + \mathbf{a}_r + \dot{\boldsymbol{\omega}} \times \mathbf{r} + \boldsymbol{\omega} \times (\boldsymbol{\omega} \times \mathbf{r}) + 2\boldsymbol{\omega} \times \mathbf{v}_r$$

Now we have

$$\ddot{\mathbf{R}}_f = -a \cos \theta\, \mathbf{i} + a \sin \theta\, \mathbf{j}$$

$$\mathbf{r} = r_0 \mathbf{i} \qquad\qquad \mathbf{v}_r = \mathbf{a}_r = 0$$

$$\boldsymbol{\omega} = \frac{V}{r_0}\mathbf{k} \qquad\qquad \dot{\boldsymbol{\omega}} = \frac{a}{r_0}\mathbf{k}$$

Substituting gives

$$\mathbf{a}_f = -a \cos \theta\, \mathbf{i} + a \sin \theta\, \mathbf{j} + a\, \mathbf{j} - \frac{v^2}{r_0}\mathbf{i}$$

$$\mathbf{a}_f = -\mathbf{i}\left[\frac{v^2}{r_0} + a \cos \theta\right] + \mathbf{j}\left(\sin \theta + 1\right)a \tag{1}$$

We want to maximize $\left|\mathbf{a}_f\right|$, or alternatively, we maximize $\left|\mathbf{a}_f\right|^2$:

$$\left|\mathbf{a}_f\right|^2 = \frac{v^4}{r_0^2} + a^2 \cos^2\theta + \frac{2av^2}{r_0}\cos\theta + a^2 + 2a^2\sin\theta + a^2\sin^2\theta$$

$$= \frac{v^4}{r_0^2} + 2a^2 + \frac{2av^2}{r_0}\cos\theta + a^2\sin^2\theta$$

$$\frac{d\left|\mathbf{a}_f\right|^2}{d\theta} = -\frac{2av^2}{r_0}\cos\theta + 2a^2\cos\theta$$

$$= 0 \text{ when } \tan\theta = \frac{ar_0}{v^2}$$

(Taking a second derivative shows this point to be a maximum.)

$$\tan\theta = \frac{ar_0}{v^2} \text{ implies } \cos\theta = \frac{v^2}{\sqrt{a^2 r_0^2 + v^4}}$$

and

$$\sin\theta = \frac{ar_0}{\sqrt{a^2 r_0^2 + v^4}}$$

Substituting into (1)

$$\mathbf{a}_f = -\mathbf{i}\left[\frac{v^2}{r_0} + \frac{av^2}{\sqrt{a^2 r_0^2 + v^4}}\right] + \mathbf{j}\left[\frac{ar_0}{\sqrt{a^2 r_0^2 + v^4}} + 1\right]a$$

This may be written as

$$\left|\mathbf{a}_f\right| = a + \sqrt{a^2 + v^4/r_0^2}$$

This is the maximum acceleration. The point which experiences this acceleration is at A:

$$\text{where } \tan\theta = \frac{ar_0}{v^2}$$

10-8.

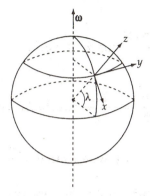

Choose the coordinates x, y, z as in the diagram. Then, the velocity of the particle and the rotation frequency of the Earth are expressed as

$$\mathbf{v} = (0, 0, \dot{z})$$
$$\boldsymbol{\omega} = \left(-\omega \cos \lambda, 0, \omega \sin \lambda\right) \qquad (1)$$

so that the acceleration due to the Coriolis force is

$$\mathbf{a} = -2\boldsymbol{\omega} \times \dot{\mathbf{r}} = 2\omega\left(0, -\dot{z} \cos \lambda, 0\right) \qquad (2)$$

This acceleration is directed along the y axis. Hence, as the particle moves along the z axis, it will be accelerated along the y axis:

$$\ddot{y} = -2\omega \dot{z} \cos \lambda \qquad (3)$$

Now, the equation of motion for the particle along the z axis is

$$\dot{z} = v_0 - gt \qquad (4)$$

$$z = v_0 t - \frac{1}{2} g t^2 \qquad (5)$$

where v_0 is the initial velocity and is equal to $\sqrt{2gh}$ if the highest point the particle can reach is h:

$$v_0 = \sqrt{2gh} \qquad (6)$$

From (3), we have

$$\dot{y} = -2\omega z \cos \lambda + c \qquad (7)$$

but the initial condition $\dot{y}(z = 0) = 0$ implies $c = 0$. Substituting (5) into (7) we find

$$\dot{y} = -2\omega \cos \lambda \left(v_0 t - \frac{1}{2} g t^2\right)$$

$$= \omega \cos \lambda \left(g t^2 - 2v_0 t^2\right) \qquad (8)$$

Integrating (8) and using the initial condition $y(t = 0) = 0$, we find

$$y = \omega \cos \lambda \left[\frac{1}{3} gt^2 - v_0 t^2 \right] \tag{9}$$

From (5), the time the particle strikes the ground ($z = 0$) is

$$0 = \left(v_0 - \frac{1}{2} gt \right) t$$

so that

$$t = \frac{2v_0}{g} \tag{10}$$

Substituting this value into (9), we have

$$y = \omega \cos \lambda \left[\frac{1}{3} g \frac{8v_0^3}{g^3} - v_0 \frac{4v_0^2}{g^2} \right]$$

$$= -\frac{4}{3} \omega \cos \lambda \frac{v_0^3}{g^2} \tag{11}$$

If we use (6), (11) becomes

$$\boxed{y = -\frac{4}{3} \omega \cos \lambda \sqrt{\frac{8h^3}{g}}} \tag{12}$$

The negative sign of the displacement shows that the particle is displaced to the *west*.

10-10.   In the previous problem we assumed the z motion to be unaffected by the Coriolis force. Actually, of course, there is an upward acceleration given by $-2\omega_x v_y$ so that

$$\ddot{z} = 2\omega V_0 \cos \alpha \cos \lambda - g \tag{1}$$

from which the time of flight is obtained by integrating twice, using the initial conditions, and then setting $z = 0$:

$$T' = \frac{2V_0 \sin \alpha}{g - 2\omega V_0 \cos \alpha \cos \lambda} \tag{2}$$

Now, the acceleration in the y direction is

$$a_y = \ddot{y} = 2\omega_x v_z$$

$$= 2(-\omega \cos \lambda)(V_0 \sin \alpha - gt) \tag{3}$$

Integrating twice and using the initial conditions, $\dot{y}(0) = V_0 \cos \alpha$ and $y(0) = 0$, we have

$$y(t) = \frac{1}{3} \omega g t^3 \cos \lambda - \omega V_0 t^2 \cos \lambda \sin \alpha + V_0 t \cos \alpha \tag{4}$$

Substituting (2) into (4), the range R′ is

$$R' = \frac{8}{3} \frac{\omega V_0^3 g \sin^3 \alpha \cos \lambda}{\left(g - 2\omega V_0 \cos \alpha \cos \lambda\right)^3} - \frac{4\omega V_0^3 \sin^3 \alpha \cos \lambda}{\left(g - 2\omega V_0 \cos \alpha \cos \lambda\right)^2} + \frac{2V_0^2 \cos \alpha \cos \lambda}{g - 2\omega V_0 \cos \alpha \cos \lambda} \tag{5}$$

We now expand each of these three terms, retaining quantities up to order ω but **neglecting all quantities proportional to ω^2 and higher powers of ω**. In the first two terms, this amounts to neglecting $2\omega V_0 \cos \alpha \cos \lambda$ compared to g in the denominator. But in the third term we must use

$$\frac{2V_0^2 \cos \alpha \sin \alpha}{g\left[1 - \dfrac{2\omega V_0}{g} \cos \alpha \cos \lambda\right]} \cong \frac{2V_0^2}{g} \cos \alpha \sin \alpha \left[1 + \frac{2\omega V_0}{g} \cos \alpha \cos \lambda\right]$$

$$= R_0' + \frac{4\omega V_0^3}{g^2} \sin \alpha \cos^2 \alpha \cos \lambda \tag{6}$$

where R_0' is the range when Coriolis effects are neglected [see Example 2.7]:

$$R_0' = \frac{2V_0^2}{g} \cos \alpha \sin \alpha \tag{7}$$

The range difference, $\Delta R' = R' - R_0'$, now becomes

$$\Delta R' = \frac{4\omega V_0^3}{g^2} \cos \lambda \left(\sin \alpha \cos^2 \alpha - \frac{1}{3} \sin^3 \alpha\right) \tag{8}$$

Substituting for V_0 in terms of R_0' from (7), we have, finally,

$$\boxed{\Delta R' = \sqrt{\frac{2R_0'}{g}} \, \omega \cos \lambda \left(\cot^{1/2} \alpha - \frac{1}{3} \tan^{3/2} \alpha\right)} \tag{9}$$

10-14. The solution to part (c) of the Problem 10-13 is modified when the particle is dropped down a mineshaft. The force due to the variation of gravity is now

$$\mathbf{a}_g \equiv -\frac{g_0}{R} \mathbf{r} + g_0 \mathbf{k} \tag{1}$$

As before, we approximate $\mathbf{r}$ for near the surface and (1) becomes

$$\mathbf{a}_g \simeq -\frac{g_0}{R}\left(x'\mathbf{i} + y'\mathbf{j} + z'\mathbf{k}\right) \tag{2}$$

In the unprimed coordinates,

$$\ddot{x} \simeq -g_0 \frac{x}{R} \tag{3}$$

To estimate the order of this term, as we probably should have done in part (c) of Problem 10-13, we can take $x \sim h^2\omega^2/g$, so that

$$\ddot{x} \sim \omega^2 h \times \frac{h}{R} \qquad\qquad (4)$$

which is reduced by a factor h/R from the accelerations obtained previously. We therefore have no southerly deflection in this order due to the variation of gravity. The Coriolis and centrifugal forces still deflect the particle, however, so that the total deflection in this approximation is

$$d \simeq \frac{3}{2} \frac{h^2}{g} \omega^2 \sin \lambda \cos \lambda \qquad\qquad (5)$$

10-19. The Coriolis force acting on the car is

$$\vec{F}_c = 2m\,\vec{v} \times \vec{\omega} \Rightarrow \left|\vec{F}_c\right| = 2mv\omega \sin \alpha$$

where $\alpha = 65°$, $m = 1300$ kg, $v = 100$ km/hr.

So $\left|\vec{F}_c\right| = 4.76$ N.

CHAPTER 11

Dynamics of Rigid Bodies

11-4.

The linear density of the rod is

$$\rho_\ell = \frac{m}{\ell} \tag{1}$$

For the origin at one end of the rod, the moment of inertia is

$$I = \int_0^\ell \rho_\ell \, x^2 \, dx = \frac{m}{\ell}\frac{\ell^3}{3} = \frac{m}{3}\ell^2 \tag{2}$$

If all of the mass were concentrated at the point which is at a distance a from the origin, the moment of inertia would be

$$I = ma^2 \tag{3}$$

Equating (2) and (3), we find

$$\boxed{a = \frac{\ell}{\sqrt{3}}} \tag{4}$$

This is the *radius of gyration*.

11-6. Let us compare the moments of inertia for the two spheres for axes through the centers of each. For the solid sphere, we have

$$I_s = \frac{2}{5} MR^2 \qquad \text{(see Problem 11-1)} \tag{1}$$

For the hollow sphere,

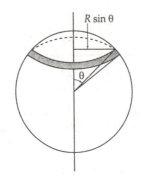

$$I_h = \sigma \int_0^{2\pi} d\phi \int_0^{\pi} \left(R \sin \theta\right)^2 R^2 \sin \theta \, d\theta$$

$$= 2\pi \sigma R^4 \int_0^{\pi} \sin^3 \theta \, d\theta$$

$$= \frac{8}{3} \pi \sigma R^4$$

or, using $4\pi \sigma R^2 = M$, we have

$$I_h = \frac{2}{3} MR^2 \tag{2}$$

Let us now roll each ball down an inclined plane. [Refer to Example 7.9.] The kinetic energy is

$$T = \frac{1}{2} M \dot{y}^2 + \frac{1}{2} I \dot{\theta}^2 \tag{3}$$

where y is the measure of the distance along the plane. The potential energy is

$$U = Mg\left(\ell - y\right) \sin \alpha \tag{4}$$

where ℓ is the length of the plane and α is the angle of inclination of the plane. Now, $y = R\theta$, so that the Lagrangian can be expressed as

$$L = \frac{1}{2} M \dot{y}^2 + \frac{1}{2} \frac{I}{R^2} \dot{y}^2 + Mgy \sin \alpha \tag{5}$$

where the constant term in U has been suppressed. The equation of motion for y is obtained in the usual way and we find

$$\ddot{y} = \frac{gMR^2 \sin \alpha}{MR^2 + I} \tag{6}$$

Therefore, the sphere with the *smaller* moment of inertia (the solid sphere) will have the *greater* acceleration down the plane.

11-11.

a) No sliding:

From energy conservation, we have

$$mg\frac{\ell}{\sqrt{2}} = mg\frac{\ell}{2} + \frac{1}{2}mv_{\text{C.M.}}^2 + \frac{1}{2}I\omega^2 \tag{1}$$

where v_{CM} is the velocity of the center of mass when one face strikes the plane; $v_{\text{C.M.}}$ is related to ω by

$$v_{\text{CM}} = \frac{\ell}{\sqrt{2}}\omega \tag{2}$$

I is the moment of inertia of the cube with respect to the axis which is perpendicular to one face and passes the center:

$$I = \frac{1}{6}m\ell^2 \tag{3}$$

Then, (1) becomes

$$\frac{mg\ell}{2}\left(\sqrt{2}-1\right) = \frac{1}{2}m\left[\frac{\ell\omega}{\sqrt{2}}\right]^2 + \frac{1}{2}\left[\frac{m\ell^2}{6}\right]\omega^2 = \frac{1}{3}m\ell^2\omega^2 \tag{4}$$

from which, we have

$$\boxed{\omega^2 = \frac{3}{2}\frac{g}{\ell}\left(\sqrt{2}-1\right)} \tag{5}$$

b) Sliding without friction:

In this case there is no external force along the horizontal direction; therefore, the cube slides so that the center of mass falls directly downward along a vertical line.

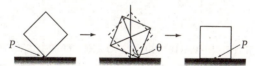

While the cube is falling, the distance between the center of mass and the plane is given by

$$y = \frac{\ell}{\sqrt{2}}\cos\theta \tag{6}$$

Therefore, the velocity of center of mass when one face strikes the plane is

$$\dot{y}\bigg|_{\theta=\pi/4} = -\frac{\ell}{\sqrt{2}}\sin\theta\,\dot{\theta}\bigg|_{\theta=\pi/4} = -\frac{1}{2}\ell\dot{\theta} = -\frac{1}{2}\ell\omega \tag{7}$$

From conservation of energy, we have

$$mg\frac{\ell}{\sqrt{2}} = mg\frac{\ell}{2} + \frac{1}{2}m\left(-\frac{1}{2}\ell\omega\right)^2 + \frac{1}{2}\left(\frac{1}{6}m\ell^2\right)\omega^2 \qquad (8)$$

from which we have

$$\boxed{\omega^2 = \frac{12}{5}\frac{g}{\ell}\left(\sqrt{2}-1\right)} \qquad (9)$$

11-14.

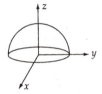

Let the surface of the hemisphere lie in the x-y plane as shown. The mass density is given by

$$\rho = \frac{M}{V} = \frac{M}{\frac{2}{3}\pi b^3} = \frac{3M}{2\pi b^3}$$

First, we calculate the center of mass of the hemisphere. By symmetry

$$x_{CM} = y_{CM} = 0$$

$$z_{CM} = \frac{1}{M}\int_v \rho z\, dv$$

Using spherical coordinates ($z = r\cos\theta,\, dv = r^2\sin\theta\, dr\, d\theta\, d\phi$) we have

$$z_{CM} = \frac{\rho}{M}\int_{\phi=0}^{2\pi} d\phi \int_{\theta=0}^{\pi/2}\sin\theta\cos\theta\, d\theta \int_{r=0}^{b} r^3\, dr$$

$$= \left[\frac{3}{2\pi b^3}\right](2\pi)\left[\frac{1}{2}\right]\left[\frac{1}{4}b^4\right] = \frac{3}{8}b$$

We now calculate the inertia tensor with respect to axes passing through the center of mass:

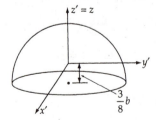

By symmetry, $I_{12} = I_{21} = I_{13} = I_{31} = I_{23} = I_{32} = 0$. Thus the axes shown are the principal axes.

Also, by symmetry $I_{11} = I_{22}$. We calculate I_{11} using Eq. 11.49:

$$I_{11} = J_{11} - M\left[\frac{3}{8}v\right]^2 \qquad (1)$$

where J_{11} = the moment of inertia with respect to the original axes

$$J_{11} = \rho \int_v \left(y^2 + z^2\right) dv$$

$$= \int_v \left(r^2 \sin^2 \theta \sin^2 \phi + r^2 \cos^2 \theta\right) r^2 \sin \theta \, dr \, d\theta \, d\phi$$

$$= \frac{3M}{2\pi b^3} \int_{r=0}^{b} r^4 \, dr \int_{\theta=0}^{\pi/2} \left[\int_{\phi=0}^{2\pi} \left(\sin^2 \theta \sin^2 \phi + \cos^2 \theta\right) d\phi\right] \sin \theta \, d\theta$$

$$= \frac{3Mb^2}{10\pi} \int_{\theta=0}^{\pi/2} \left(\pi \sin^3 \theta + 2\pi \cos^2 \theta \sin \theta\right) d\theta$$

$$= \frac{2}{5} Mb^2$$

Thus, from (1)

$$I_{11} = I_{22} = \frac{2}{5} Mb^2 - \frac{9}{64} Mb^2 = \frac{83}{320} Mb^2$$

Also, from Eq. 11.49

$$I_{33} = J_{33} - M(0) = J_{33}$$

($I_{33} = J_{33}$ should be obvious physically)

So

$$I_{33} = \rho \int_v \left(x^2 + y^2\right) dv$$

$$= \rho \int_v r^4 \sin^3 \theta \, dr \, d\theta \, d\phi = \frac{2}{5} Mb^2$$

> Thus, the principal axes are the primed axes shown in the figure. The principal moments of inertia are
>
> $$I_{11} = I_{22} = \frac{83}{320} Mb^2$$
>
> $$I_{33} = \frac{2}{5} Mb^2$$

11-17.

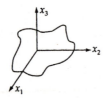

The plate is assumed to have negligible thickness and the mass per unit area is ρ_s. Then, the inertia tensor elements are

$$I_{11} = \rho_s \int \left(r^2 - x_1^2 \right) dx_1 \, dx_2$$

$$= \rho_s \int \left(x_2^2 + x_3^2 \right) dx_1 dx_2 = \rho_s \int x_2^2 \, dx_1 \, dx_2 \equiv A \tag{1}$$

$$I_{22} = \rho_s \int \left(r^2 - x_2^2 \right) dx_1 \, dx_2 = \rho_s \int x_1^2 \, dx_1 \, dx_2 \equiv B \tag{2}$$

$$I_{33} = \rho_s \int \left(r^2 - x_3^2 \right) dx_1 \, dx_2 = \rho_s \int \left(x_1^2 + x_2^2 \right) dx_1 \, dx_2 \tag{3}$$

Defining A and B as above, I_{33} becomes

$$I_{33} = A + B \tag{4}$$

Also,

$$I_{12} = \rho_s \int \left(-x_1 x_2 \right) dx_1 \, dx_2 \equiv -C \tag{5}$$

$$I_{21} = \rho_s \int \left(-x_2 x_1 \right) dx_1 \, dx_2 = -C \tag{6}$$

$$I_{13} = \rho_s \int \left(-x_1 x_3 \right) dx_1 \, dx_2 = 0 = I_{31} \tag{7}$$

$$I_{23} = \rho_s \int \left(-x_2 x_3 \right) dx_1 \, dx_2 = 0 = I_{32} \tag{8}$$

Therefore, the inertia tensor has the form

$$\{\mathbf{I}\} = \begin{bmatrix} A & -C & 0 \\ -C & B & 0 \\ 0 & 0 & A+B \end{bmatrix} \tag{9}$$

11-22. According to Eq. (11.61),

$$I'_{ij} = \sum_{k,l} \lambda_{ik} \, I_{k\ell} \, \lambda_{\ell j}^{-1} \tag{1}$$

Then,

$$tr\{\mathbf{I'}\} = \sum_i I'_{ii} = \sum_i \sum_{k,\ell} \lambda_{ik}\, I_{k\ell}\, \lambda_{\ell i}^{-1}$$

$$= \sum_{k,\ell} I_{k\ell} \sum_i \lambda_{\ell i}^{-1}\, \lambda_{ik}$$

$$= \sum_{k,\ell} I_{k\ell}\, \delta_{\ell k} = \sum_k I_{kk} \tag{2}$$

so that

$$\boxed{tr\{\mathbf{I'}\} = tr\{\mathbf{I}\}} \tag{3}$$

This relation can be verified for the examples in the text by straightforward calculations.

Note: A *translational* transformation is *not* a *similarity* transformation and, in general, $tr\{\mathbf{I}\}$ is not invariant under translation. (For example, $tr\{\mathbf{I}\}$ will be different for inertia tensors expressed in coordinate system with different origins.)

11-25.

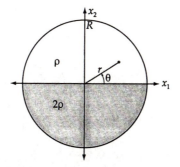

The center of mass of the disk is $(0, \overline{x}_2)$, where

$$\overline{x}_2 = \frac{\rho}{M}\left[2 \int_{\substack{\text{lower} \\ \text{semicircle}}} x_2\, dx_1\, dx_2 + \int_{\substack{\text{upper} \\ \text{semicircle}}} x_2\, dx_1\, dx_2 \right]$$

$$= \frac{\rho}{M}\left[\int_0^R \int_0^\pi (r\sin\theta)\cdot r\,dr\,d\theta + 2\int_0^R \int_\pi^{2\pi} (r\sin\theta)\cdot r\,dr\,d\theta \right]$$

$$= -\frac{2}{3}\frac{\rho R^3}{M} \tag{1}$$

Now, the mass of the disk is

$$M = \rho \cdot \frac{1}{2}\pi R^2 + 2\rho \cdot \frac{1}{2}\pi R^2$$

$$= \frac{3}{2}\rho \pi R^2 \tag{2}$$

so that

$$\bar{x}_2 = -\frac{4}{9\pi} R \tag{3}$$

The direct calculation of the rotational inertia with respect to an axis through the center of mass is tedious, so we first compute I with respect to the x_3-axis and then use Steiner's theorem.

$$I_3 = \rho \left[\int_0^R \int_0^\pi r^2 \cdot r dr\, d\theta + 2 \int_0^R \int_\pi^{2\pi} r^2 \cdot r dr\, d\theta \right]$$

$$= \frac{3}{4} \pi \rho R^4 = \frac{1}{2} MR^2 \tag{4}$$

Then,

$$I_0 = I_3 - M\bar{x}_2^2$$

$$= \frac{1}{2} MR^2 - M \cdot \frac{16}{81\pi^2} R^2$$

$$= \frac{1}{2} MR^2 \left[1 - \frac{32}{81\pi^2} \right] \tag{5}$$

When the disk rolls without slipping, the velocity of the center of mass can be obtained as follows:

Thus

$$x_{CM} = R\theta - \left|\bar{x}_2\right| \sin \theta$$

$$y_{CM} = R - \left|\bar{x}_2\right| \cos \theta$$

$$\dot{x}_{CM} = R\dot\theta - \left|\bar{x}_2\right| \dot\theta \cos \theta$$

$$\dot{y}_{CM} = \left|\bar{x}_2\right| \dot\theta \sin \theta$$

$$\left(\dot{x}_{CM}^2 + \dot{y}_{CM}^2 \right) = V^2 = R^2\dot\theta^2 + \bar{x}_2^2\, \dot\theta^2 - 2\dot\theta^2\, R\left|\bar{x}_2\right| \cos \theta$$

$$V^2 = a^2\dot\theta^2 \tag{6}$$

where

$$a = \sqrt{R^2 + \bar{x}_2^2 - 2R\left|\bar{x}_2\right| \cos \theta} \tag{7}$$

Using (3), a can be written as

$$a = R\sqrt{1 + \frac{16}{81\pi^2} - \frac{8}{9\pi}\cos\theta} \tag{8}$$

The kinetic energy is

$$T = T_{\text{trans}} + T_{\text{rot}}$$

$$= \frac{1}{2}Mv^2 + \frac{1}{2}I_0\dot{\theta}^2 \tag{9}$$

Substituting and simplifying yields

$$T = \frac{1}{2}MR^2\dot{\theta}^2\left[\frac{3}{2} - \frac{8}{9\pi}\cos\theta\right] \tag{10}$$

The potential energy is

$$U = Mg\left[\frac{1}{2}R + \bar{x}_2\cos\theta\right]$$

$$= \frac{1}{2}MgR\left[1 - \frac{8}{9\pi}\cos\theta\right] \tag{11}$$

Thus the Lagrangian is

$$\boxed{L = \frac{1}{2}MR\left[R\dot{\theta}^2\left[\frac{3}{2} - \frac{8}{9\pi}\cos\theta\right] - g\left[1 - \frac{8}{9\pi}\cos\theta\right]\right]} \tag{12}$$

11-31. The moments of inertia of the plate are

$$\left.\begin{array}{l} I_1 = I_2\cos 2\alpha \\[2mm] I_2 \\[2mm] I_3 = I_1 + I_2 \\[2mm] \quad = I_2(1 + \cos 2\alpha) \\[2mm] \quad = 2I_2\cos^2\alpha \end{array}\right\} \tag{1}$$

We also note that

$$I_1 - I_2 = -I_2(1 - \cos 2\alpha)$$

$$= -2I_2\sin^2\alpha \tag{2}$$

Since the plate moves in a force-free manner, the Euler equations are [see Eq. (11.114)]

$$\left.\begin{array}{l} \left(I_1 - I_2\right)\omega_1\omega_2 - I_3\dot{\omega}_3 = 0 \\[1ex] \left(I_2 - I_3\right)\omega_2\omega_3 - I_1\dot{\omega}_1 = 0 \\[1ex] \left(I_3 - I_1\right)\omega_3\omega_1 - I_2\dot{\omega}_2 = 0 \end{array}\right\} \tag{3}$$

Substituting (1) and (2) into (3), we find

$$\left.\begin{array}{l} \left(-2I_2 \sin^2 \alpha\right)\omega_1\omega_2 - \left(2I_2 \cos^2 \alpha\right)\dot{\omega}_3 = 0 \\[1ex] \left(-I_2 \cos 2\alpha\right)\omega_2\omega_3 - \left(I_2 \cos 2\alpha\right)\dot{\omega}_1 = 0 \\[1ex] I_2\,\omega_3\omega_1 - I_2\dot{\omega}_2 = 0 \end{array}\right\} \tag{4}$$

These equations simplify to

$$\left.\begin{array}{l} \dot{\omega}_3 = -\omega_1\omega_2 \tan^2 \alpha \\[1ex] \dot{\omega}_1 = -\omega_2\omega_3 \\[1ex] \dot{\omega}_2 = \omega_3\omega_1 \end{array}\right\} \tag{5}$$

From which we can write

$$\omega_1\omega_2\omega_3 = \omega_2\dot{\omega}_2 = -\omega_1\dot{\omega}_1 = -\omega_3\dot{\omega}_3 \cot^2 \alpha \tag{6}$$

Integrating, we find

$$\omega_2^2 - \omega_2^2(0) = -\omega_1^2 + \omega_1^2(0) = -\omega_3^2 \cot^2 \alpha + \omega_3^2(0) \cot^2 \alpha \tag{7}$$

Now, the initial conditions are

$$\left.\begin{array}{l} \omega_1(0) = \Omega \cos \alpha \\[1ex] \omega_2(0) = 0 \\[1ex] \omega_3(0) = \Omega \sin \alpha \end{array}\right\} \tag{8}$$

Therefore, the equations in (7) become

$$\omega_2^2 = -\omega_1^2 + \Omega^2 \cos^2 \alpha = -\omega_3^2 \cot^2 \alpha + \Omega^2 \cos^{2\alpha} \tag{9}$$

From (5), we can write

$$\dot{\omega}_2^2 = \omega_3^2\,\omega_1^2 \tag{10}$$

and from (9), we have $\omega_1^2 = \omega_3^2 \cot^2 \alpha$. Therefore, (10) becomes

$$\dot{\omega}_2 = \omega_3^2 \cot \alpha \tag{11}$$

and using $\omega_3^2 = \Omega^2 \sin^2 \alpha - \omega_2^2 \tan^2 \alpha$ from (9), we can write (11) as

$$\frac{\dot{\omega}_2}{\omega_2^2 \tan^2 \alpha - \Omega^2 \sin^2 \alpha} = -\cot \alpha \tag{12}$$

Since $\dot{\omega}_2 = d\omega_2/dt$, we can express this equation in terms of integrals as

$$\int \frac{d\omega_2}{\omega_2^2 \tan^2 \alpha - \Omega^2 \sin^2 \alpha} = -\cot \alpha \int dt \tag{13}$$

Using Eq. (E.4c), Appendix E, we find

$$-\frac{1}{(\tan \alpha)(\Omega \sin \alpha)} \tanh^{-1}\left[\frac{\omega_2 \tan \alpha}{\Omega \sin \alpha}\right] = -t \cot \alpha \tag{14}$$

Solving for ω_2,

$$\boxed{\omega_2(t) = \Omega \cos \alpha \tanh(\Omega t \sin \alpha)} \tag{15}$$

11-34. The Euler equation, which describes the rotation of an object about its symmetry axis, say $0x$, is

$$I_x \dot{\omega}_x - (I_y - I_z)\omega_y\omega_z = N_x$$

where $N_x = -b\,\omega_x$ is the component of torque along Ox. Because the object is symmetric about Ox, we have $I_y = I_z$, and the above equation becomes

$$I_x \frac{d\omega_x}{dt} = -b\,\omega_x \quad \Rightarrow \quad \omega_x = e^{-\frac{b}{I_x}t}\,\omega_{x0}$$

Coupled Oscillations

12-3. The equations of motion are

$$\ddot{x}_1 + \frac{m}{M}\ddot{x}_2 + \omega_0^2\, x_1 = 0 \left.\vphantom{\begin{array}{c}1\\1\end{array}}\right]$$

$$\ddot{x}_2 + \frac{m}{M}\ddot{x}_2 + \omega_0^2\, x_2 = 0 \qquad\qquad (1)$$

We try solutions of the form

$$x_1(t) = B_1\, e^{i\omega t}; \quad x_2(t) = B_2\, e^{i\omega t} \qquad\qquad (2)$$

We require a non-trivial solution (i.e., the determinant of the coefficients of B_1 and B_2 equal to zero), and obtain

$$\left(\omega_0^2 - \omega^2\right)^2 - \omega^4\left[\frac{m}{M}\right]^2 = 0 \qquad\qquad (3)$$

so that

$$\omega_0^2 - \omega^2 = \pm\omega^2\,\frac{m}{M} \qquad\qquad (4)$$

and then

$$\omega^2 = \frac{\omega_0^2}{1 \pm \dfrac{m}{M}} \qquad\qquad (5)$$

Therefore, the frequencies of the normal modes are

$$\omega_1 = \sqrt{\frac{\omega_0^2}{1 + \frac{m}{M}}}$$

$$\omega_2 = \sqrt{\frac{\omega_0^2}{1 - \frac{m}{M}}}$$

(6)

where ω_1 corresponds to the symmetric mode and ω_2 to the antisymmetric mode.

By inspection, one can see that the normal coordinates for this problem are the same as those for the example of Section 12.2 [i.e., Eq. (12.11)].

12-6.

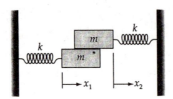

If the frictional force acting on mass 1 due to mass 2 is

$$f = -\beta\left(\dot{x}_1 - \dot{x}_2\right)$$

(1)

then the equations of motion are

$$m\ddot{x}_1 + \beta\left(\dot{x}_1 - \dot{x}_2\right) + \kappa x_1 = 0$$

$$m\ddot{x}_2 + \beta\left(\dot{x}_2 - \dot{x}_1\right) + \kappa x_2 = 0$$

(2)

Since the system is not conservative, the eigenfrequencies will not be entirely real as in the previous cases. Therefore, we attempt a solution of the form

$$x_1(t) = B_1 e^{\alpha t}; \quad x_2(t) = B_2 e^{\alpha t}$$

(3)

where $\alpha = \lambda + i\omega$ is a complex quantity to be determined. Substituting (3) into (1), we obtain the following secular equation by setting the determinant of the coefficients of the B's equal to zero:

$$\left(m\alpha^2 + \beta\alpha + \kappa\right)^2 = \beta^2\alpha^2$$

(4)

from which we find the two solutions

$$\alpha_1 = \pm i\sqrt{\frac{\kappa}{m}}; \quad \omega_1 = \pm\sqrt{\frac{\kappa}{m}}$$

$$\alpha_2 = \frac{1}{m}\left(-\beta \pm \sqrt{\beta^2 - m\kappa}\right)$$

(5)

The general solution is therefore

$$x_1(t) = B_{11}^+ \, e^{i\sqrt{\kappa/m}\,t} + B_{11}^- \, e^{i\sqrt{\kappa/m}\,t} + e^{-\beta t/m}\left(B_{12}^+ \, e^{\sqrt{\beta^2 - m\kappa}\,t/m} + B_{12}^- \, e^{\sqrt{\beta^2 - m\kappa}\,t/m} \right) \qquad (6)$$

and similarly for $x_2(t)$.

The first two terms in the expression for $x_1(t)$ are purely oscillatory, whereas the last two terms contain the damping factor $e^{-\beta t}$. (Notice that the term $B_{12}^+ \exp\left(\sqrt{\beta^2 - m\kappa}\, t \right)$ *increases* with time if $\beta^2 > m\kappa$, but B_{12}^+ is not required to vanish in order to produce physically realizable motion because the damping term, $\exp(-\beta t)$, *decreases* with time at a more rapid rate; that is $-\beta + \sqrt{\beta^2 - m\kappa} < 0$.)

To what modes do α_1 and α_2 apply? In Mode 1 there is purely oscillating motion without friction. This can happen only if the two masses have no relative motion. Thus, Mode 1 is the *symmetric* mode in which the masses move *in phase*. Mode 2 is the *antisymmetric* mode in which the masses move *out of phase* and produce frictional damping. If $\beta^2 < m\kappa$, the motion is one of damped oscillations, whereas if $\beta^2 > m\kappa$, the motion proceeds monotonically to zero amplitude.

12-11. Taking a time derivative of the equations gives $(\dot{q} = I)$

$$L\ddot{I}_1 + \frac{I_1}{C} + M\ddot{I}_2 = 0$$

$$L\ddot{I}_2 + \frac{I_2}{C} + M\ddot{I}_1 = 0$$

Assume $I_1 = B_1 e^{i\omega t}$, $I_2 = B_2 e^{i\omega t}$; and substitute into the previous equations. The result is

$$-\omega^2 L B_1 e^{i\omega t} + \frac{1}{C} B_1 e^{i\omega t} - M\omega^2 B_2 e^{i\omega t} = 0$$

$$-\omega^2 L B_2 e^{i\omega t} + \frac{1}{C} B_2 e^{i\omega t} - M\omega^2 B_1 e^{i\omega t} = 0$$

These reduce to

$$B_1 \left[\frac{1}{C} - \omega^2 L \right] + B_2 \left(-M\omega^2 \right) = 0$$

$$B_1 \left(-M\omega^2 \right) + B_2 \left[\frac{1}{C} - \omega^2 L \right] = 0$$

This implies that the determinant of coefficients of B_1 and B_2 must vanish (for a non-trivial solution). Thus

$$\begin{vmatrix} \dfrac{1}{C} - \omega^2 L & -M\omega^2 \\[2ex] -M\omega^2 & \dfrac{1}{C} - \omega^2 L \end{vmatrix} = 0$$

$$\left[\frac{1}{C} - \omega^2 L\right]^2 - \left(M\omega^2\right)^2 = 0$$

$$\frac{1}{C} - \omega^2 L = \pm M\omega^2$$

or

$$\omega^2 = \frac{1}{C(L \pm M)}$$

Thus

$$\omega_1 = \sqrt{\frac{1}{C(L+M)}}$$

$$\omega_2 = \sqrt{\frac{1}{C(L-M)}}$$

12-14.

The Kirchhoff circuit equations are (after differentiating and using $\dot{q} = I$)

$$L_1 \ddot{I}_1 + \left[\frac{1}{C_1} + \frac{1}{C_{12}}\right] I_1 - \frac{1}{C_{12}} I_2 = 0$$

$$L_2 \ddot{I}_2 + \left[\frac{1}{C_2} + \frac{1}{C_{12}}\right] I_2 - \frac{1}{C_{12}} I_1 = 0$$

(1)

Using a harmonic time dependence for $I_1(t)$ and $I_2(t)$, the secular equation is found to be

$$\left[L_1\omega^2 - \frac{C_1 + C_{12}}{C_1 C_{12}}\right]\left[L_2\omega^2 - \frac{C_2 + C_{12}}{C_2 C_{12}}\right] = \frac{1}{C_{12}^2}$$

(2)

Solving for the frequency,

$$\omega^2 = \frac{C_1 L_1 (C_2 + C_{12}) + C_2 L_2 (C_1 + C_{12}) \pm \sqrt{\left[C_1 L_1 (C_2 + C_{12}) - C_2 L_2 (C_1 + C_{12})\right]^2 + 4C_1^2 C_2^2 L_1 L_2}}{2L_1 L_2 C_1 C_2 C_{12}}$$

(3)

Because the characteristic frequencies are given by this complicated expression, we examine the normal modes for the special case in which $L_1 = L_2 = L$ and $C_1 = C_2 = C$. Then,

$$\boxed{\begin{aligned} \omega_1^2 &= \frac{2C + C_{12}}{LCC_{12}} \\[2mm] \omega_2^2 &= \frac{1}{LC} \end{aligned}}$$

(4)

Observe that ω_2 corresponds to the case of uncoupled oscillations. The equations for this simplified circuit can be set in the same form as Eq. (12.1), and consequently the normal modes can be found in the same way as in Section 12.2. There will be two possible modes of oscillation: (1) *out of phase*, with frequency ω_1, and (2) *in phase*, with frequency ω_2.

Mode 1 corresponds to the currents I_1 and I_2 oscillating always *out of phase*:

Mode 2 corresponds to the currents I_1 and I_2 oscillating always *in phase*:

(The analogy with two oscillators coupled by a spring can be seen by associating case 1 with Fig. 12-2 for $\omega = \omega_1$ and case 2 with Fig. 12-2 for $\omega = \omega_2$.) If we now let $L_1 \neq L_2$ and $C_1 \neq C_2$, we do not have pure symmetrical and antisymmetrical symmetrical modes, but we can associate ω_2 with the mode of highest degree of symmetry and ω_1 with that of lowest degree of symmetry.

12-19.   With the given expression for U, we see that $\{A\}$ has the form

$$\{\mathbf{A}\} = \begin{bmatrix} 1 & -\varepsilon_{12} & -\varepsilon_{13} \\ -\varepsilon_{12} & 1 & -\varepsilon_{23} \\ -\varepsilon_{13} & -\varepsilon_{23} & 1 \end{bmatrix}$$

(1)

The kinetic energy is

$$T = \frac{1}{2}\left(\dot{\theta}_1^2 + \dot{\theta}_2^2 + \dot{\theta}_3^2 \right)$$

(2)

so that $\{\mathbf{m}\}$ is

$$\{\mathbf{m}\} = \begin{bmatrix} 1 & 0 & 0 \\ 0 & 1 & 0 \\ 0 & 0 & 1 \end{bmatrix}$$

(3)

The secular determinant is

$$\begin{vmatrix} 1-\omega^2 & -\varepsilon_{12} & -\varepsilon_{13} \\ -\varepsilon_{12} & 1-\omega^2 & -\varepsilon_{23} \\ -\varepsilon_{13} & -\varepsilon_{23} & 1-\omega^2 \end{vmatrix} = 0 \tag{4}$$

Thus,

$$\left(1-\omega^2\right)^3 - \left(1-\omega^2\right)\left(\varepsilon_{12}^2 + \varepsilon_{13}^2 + \varepsilon_{23}^2\right) - 2\varepsilon_{12}\varepsilon_{13}\varepsilon_{23} = 0 \tag{5}$$

This equation is of the form (with $1-\omega^2 \equiv x$)

$$x^3 - 3\alpha^2 x - 2\beta^2 = 0 \tag{6}$$

which has a double root *if and only if*

$$\left(\alpha^2\right)^{3/2} = \beta^2 \tag{7}$$

Therefore, (5) will have a double root *if and only if*

$$\left[\frac{\varepsilon_{12}^2 + \varepsilon_{13}^2 + \varepsilon_{23}^2}{3}\right]^{3/2} = \varepsilon_{12}\varepsilon_{13}\varepsilon_{23} \tag{8}$$

This equation is satisfied only if

$$\boxed{\varepsilon_{12} = \varepsilon_{13} = \varepsilon_{23}} \tag{9}$$

Consequently, there will be no degeneracy unless the three coupling coefficients are identical.

12-23.   The total energy of the r-th normal mode is

$$E_r = T_r + U_r$$
$$\tag{1}$$
$$= \frac{1}{2}\dot{\eta}_r^2 + \frac{1}{2}\omega_r^2\eta_r^2$$

where

$$\eta_r = \beta_r e^{i\omega_r t} \tag{2}$$

Thus,

$$\dot{\eta}_r = i\omega_r \beta_r e^{i\omega_r t} \tag{3}$$

In order to calculate T_r and U_r, we must take the squares of the real parts of $\dot{\eta}_r$ and η_r:

$$\dot{\eta}_r^2 = \left(\mathrm{Re}\,\dot{\eta}_r\right)^2 = \left[\mathrm{Re}\,i\,\omega_r\left(\mu_r + i\nu_r\right)\left(\cos\omega_r t + i\sin\omega_r t\right)\right]^2$$

$$= \left[-\omega_r\,\nu_r\cos\omega_r t - \omega_r\,\mu_r\sin\omega_r t\right]^2 \tag{4}$$

so that

$$T_r = \frac{1}{2}\omega_r^2\left[\nu_r\cos\omega_r t + \mu_r\sin\omega_r t\right]^2 \tag{5}$$

Also

$$\eta_r^2 = \left(\operatorname{Re}\eta_r\right)^2 = \left[\operatorname{Re}\left(\mu_r + i\nu_r\right)\left(\cos\omega_r t + i\sin\omega_r t\right)\right]^2$$

$$= \left[\mu_r \cos\omega_r t - \nu_r \sin\omega_r t\right]^2 \qquad (6)$$

so that

$$U_r = \frac{1}{2}\omega_r^2 \left[\mu_r \cos\omega_r t - \nu_r \, xin\, \omega_r t\right]^2 \qquad (7)$$

Expanding the squares in T_r and U_r, and then adding, we find

$$E_r = T_r + U_r$$

$$= \frac{1}{2}\omega_r^2 \left(\mu_r^2 + \nu_r^2\right)$$

Thus,

$$\boxed{E_r = \frac{1}{2}\omega_r^2 |\beta_r|^2} \qquad (8)$$

So that the total energy associated with each normal mode is separately conserved.

For the case of Example 12.3, we have for Mode 1

$$\eta_1 = \sqrt{\frac{M}{2}}\left(x_{10} - x_{20}\right)\cos\omega_1 t \qquad (9)$$

Thus,

$$\dot{\eta}_1 = -\omega_1 \sqrt{\frac{M}{2}}\left(x_{10} - x_{20}\right)\sin\omega_1 t \qquad (10)$$

Therefore,

$$E_1 = \frac{1}{2}\dot{\eta}_1^2 + \frac{1}{2}\omega_1^2 \eta_1^2 \qquad (11)$$

But

$$\omega_1^2 = \frac{\kappa + 2\kappa_{12}}{M} \qquad (12)$$

so that

$$E_1 = \frac{1}{2}\frac{\kappa + 2\kappa_{12}}{M}\frac{M}{2}\left(x_{10} - x_{20}\right)^2 \sin^2\omega_1 t + \frac{1}{2}\frac{\kappa + 2\kappa_{12}}{M}\frac{M}{2}\left(x_{10} - x_{20}\right)^2 \cos^2\omega_1 t$$

$$= \frac{1}{4}\left(\kappa + 2\kappa_{12}\right)\left(x_{10} - x_{20}\right)^2 \qquad (13)$$

which is recognized as the value of the potential energy at $t = 0$. [At $t = 0$, $\dot{x}_1 = \dot{x}_2 = 0$, so that the total energy is $U_1(t = 0)$.]

12-27. The coordinates of the system are given in the figure:

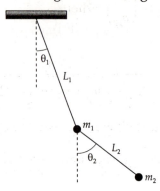

Kinetic energy:

$$T = \frac{1}{2}m_1^2\,\dot{\theta}_1^2\,L_1^2 + \frac{1}{2}m_2\left(\dot{L}_1^2\,\dot{\theta}_1^2 + L_2^2\,\dot{\theta}_2^2 - 2L_1L_2\dot{\theta}_1\dot{\theta}_2\cos(\theta_1 - \theta_2)\right)$$

$$\approx \frac{1}{2}\dot{\theta}_1^2\left(m_1L_1^2 + m_2L_1^2\right) + \frac{1}{2}m_2L_2^2\dot{\theta}_2^2 - m_2L_1L_2\dot{\theta}_1\dot{\theta}_2 = \frac{1}{2}\sum_{jk}m_{jk}\,\dot{\theta}_j\,\dot{\theta}_k$$

$$\Rightarrow \left[m_{jk}\right] = \begin{bmatrix} (m_1 + m_2)L_1^2 & -m_2L_1L_2 \\ -m_2L_1L_2 & m_2L_2^2 \end{bmatrix}$$

Potential energy:

$$U = m_1gL_1\left(1 - \cos\theta_1\right) + m_2g\left[L_1\left(1 - \cos\theta_1\right) + L_2\left(1 - \cos\theta_2\right)\right]$$

$$\approx \left(m_1 + m_2\right)gL_1\frac{\theta_1^2}{2} + m_2gL_2\frac{\theta_2^2}{2} = \frac{1}{2}\sum_{jk}A_{jk}\theta_j\theta_k$$

$$\Rightarrow \left[A_{jk}\right] = \begin{bmatrix} (m_1 + m_2)gL_1 & 0 \\ 0 & m_2gL_2 \end{bmatrix}$$

Proper oscillation frequencies are solutions of the equation

$$\mathrm{Det}\left([A] - \omega^2[m]\right) = 0$$

$$\Rightarrow \quad \omega_{1,2} = \frac{(m_1 + m_2)g(L_1 + L_2) \pm \sqrt{(m_1 + m_2)g^2\left[m_1(L_1 - L_2)^2 + m_2(L_1 + L_2)^2\right]}}{2m_1L_1L_2}$$

The eigenstate corresponding to ω_1 is $\begin{pmatrix} a_{11} \\ a_{21} \end{pmatrix}$ where

$$a_{21} = \frac{(m_1 + m_2)L_1}{m_1L_2}\left(1 - \frac{2gm_1L_2}{(m_1 + m_2)g(L_1 + L_2) + \sqrt{(m_1 + m_2)g^2\left[m_1(L_1 - L_2)^2 + m_2(L_1 + L_2)^2\right]}}\right) \times a_{11}$$

The eigenstate corresponding to ω_2 is $\begin{pmatrix} a_{12} \\ a_{22} \end{pmatrix}$ where

$$a_{22} = \frac{(m_1 + m_2)L_1}{m_1 L_2} \left(1 - \frac{2gm_1 L_2}{(m_1 + m_2)g(L_1 + L_2) - \sqrt{(m_1 + m_2)g^2\left[m_1(L_1 - L_2)^2 + m_2(L_1 + L_2)^2\right]}} \right) \times a_{12}$$

These expressions are rather complicated; we just need to note that a_{11} and a_{21} have the same sign $\left(\dfrac{a_{11}}{a_{21}} > 0 \right)$ while a_{12} and a_{22} have opposite sign $\left(\dfrac{a_{11}}{a_{21}} < 0 \right)$.

The relationship between coordinates (θ_1, θ_2) and normal coordinates η_1, η_2 are

$$\left. \begin{aligned} \theta_1 &= a_{11}\,\eta_1 + a_{12}\,\eta_2 \\ \theta_2 &= a_{21}\,\eta_1 + a_{22}\,\eta_2 \end{aligned} \right\} \qquad \Leftrightarrow \qquad \begin{cases} \eta_1 \sim \theta_1 - \dfrac{a_{12}}{a_{22}}\theta_2 \\[2mm] \eta_2 \sim \theta_1 - \dfrac{a_{11}}{a_{21}}\theta_2 \end{cases}$$

To visualize the normal coordinate η_1, let $\eta_2 = 0$. Then to visualize the normal coordinate η_2, we let $\eta_1 = 0$. Because $\dfrac{a_{11}}{a_{21}} > 0$ and $\dfrac{a_{12}}{a_{22}} < 0$, we see that these normal coordinates describe two oscillation modes. In the first one, the two bobs move in opposite directions and in the second, the two bobs move in the same direction.

CHAPTER 13

Continuous Systems; Waves

13-4. The coefficients ν_r are all zero and the μ_r are given by Eq. (13.8a):

$$\mu_r = \frac{8}{L^2} \int_0^L x(L-x) \sin \frac{r\pi x}{L} \, dx$$

$$= \frac{16}{r^3 \pi^3} \left[1 - (-1)^r \right] \tag{1}$$

so that

$$\mu_r = \left[\begin{array}{ll} 0, & r \text{ even} \\[2mm] \dfrac{32}{r^3 \pi \sqrt{3}}, & r \text{ odd} \end{array} \right. \tag{2}$$

Since

$$q(x,t) = \sum_r \mu_r \sin \frac{r\pi x}{L} \cos \omega_r t \tag{3}$$

the amplitude of the n-th mode is just μ_n.

The characteristic frequencies are given by Eq. (13.11):

$$\omega_n = \frac{n\pi}{L} \sqrt{\frac{\tau}{\rho}} \tag{4}$$

13-7.

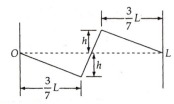

Since $\dot{q}(x,0)=0$, we know that all of the v_r are zero and the μ_r are given by Eq. (13.8a):

$$\mu_r = \frac{2}{L}\int_0^L q(x,0)\sin\frac{r\pi x}{L}\,dx \tag{1}$$

The initial condition on $q(x,t)$ is

$$q(x,0) = \begin{bmatrix} -\dfrac{7h}{3L}x, & 0\le x\le\dfrac{3}{7}L \\[2ex] \dfrac{7h}{L}(2x-L), & \dfrac{3}{7}L\le x\le\dfrac{4}{7}L \\[2ex] \dfrac{7h}{3L}(L-x), & \dfrac{4}{7}L\le x\le L \end{bmatrix} \tag{2}$$

Evaluating the μ_r we find

$$\mu_r = \frac{98}{3}\frac{h}{r^2\pi^2}\left[\sin\frac{4r\pi}{7}-\sin\frac{3r\pi}{7}\right] \tag{3}$$

Obviously, $\mu_r = 0$ when $4r/7$ and $3r/7$ simultaneously are integers. This will occur when r is any multiple of 7 and so we conclude that the modes with frequencies that are multiples of $7\omega_1$ will be absent.

13-12. The equation to be solved is

$$\ddot{\eta}_s + \frac{D}{\rho}\dot{\eta}_s + \frac{s^2\pi^2\tau}{\rho b}\eta_s = 0 \tag{1}$$

Compare this equation to Eq. (3.35):

$$\ddot{x} + 2\beta\dot{x} + \omega_0^2\,x = 0$$

The solution to Eq. (3.35) is Eq. (3.37):

$$x(t) = e^{-\beta t}\left[A_1\exp\!\left(\sqrt{\beta^2-\omega_0^2}\;t\right) + A_2\exp\!\left(-\sqrt{\beta^2-\omega_0^2}\;t\right)\right]$$

Thus, by analogy, the solution to (1) is

$$n_s(t) = e^{-Dt/2\rho}\left[A_1\exp\!\left[\sqrt{\frac{D^2}{4\rho^2}-\frac{s^2\pi^2\tau}{\rho b}}\;t\right] + A_2\exp\!\left[-\sqrt{\frac{D^2}{4\rho^2}-\frac{s^2\pi^2\tau}{\rho b}}\;t\right]\right]$$

13-13. Assuming k is real, while ω and v are complex, the wave function becomes

$$\psi(x,t) = Ae^{i(\alpha t + i\beta t - kx)}$$

$$= Ae^{(\alpha t - kx)} e^{-\beta t} \tag{1}$$

whose real part is

$$\psi(x,t) = Ae^{-\beta t} \cos(\alpha t - kx) \tag{2}$$

and the wave is damped in time, with damping coefficient β.

From the relation

$$k^2 = \frac{\omega^2}{v^2} \tag{3}$$

we obtain

$$(\alpha + i\beta)^2 = k^2 (u + iw)^2 \tag{4}$$

By equating the real and imaginary part of this equation we can solve for α and β in terms of u and w:

$$\alpha = \frac{k^2 uw}{\beta} \tag{5}$$

and

$$\beta = \begin{bmatrix} kw \\ iku \end{bmatrix} \tag{6}$$

Since we have assumed β to be real, we choose the solution

$$\boxed{\beta = kw} \tag{7}$$

Substituting this into (5), we have

$$\boxed{\alpha = ku} \tag{8}$$

as expected.

Then, the phase velocity is obtained from the oscillatory factor in (2) by its definition:

$$V = \frac{\mathrm{Re}\,\omega}{k} = \frac{\alpha}{k} \tag{9}$$

That is,

$$\boxed{V = u}$$

13-17.   We let

$$m_j = \begin{bmatrix} m', & j = 2n \\ m'', & j = 2n+1 \end{bmatrix} \tag{1}$$

where n is an integer.

Following the procedure in Section 12.9, we write

$$F_{2n} = m'\ddot{q}_{2n} = \frac{\tau}{d}\left(q_{2n-1} - 2q_{2n} + q_{2n+1}\right) \tag{2a}$$

$$F_{2n+1} = m''\ddot{q}_{2n+1} = \frac{\tau}{d}\left(q_{2n} - 2q_{2n+1} + q_{2n+2}\right) \tag{2b}$$

Assume solutions of the form

$$q_{2n} = Ae^{i(\omega t - 2nkd)} \tag{3a}$$

$$q_{2n+1} = Be^{i[\omega t - (2n+1)kd]} \tag{3b}$$

Substituting (3a,b) into (2a,b), we obtain

$$\left. \begin{aligned} -\omega^2 A &= \frac{\tau}{m'd}\left(Be^{ikd} - 2A + Be^{-ikd}\right) \\ -\omega^2 B &= \frac{\tau}{m''d}\left(Ae^{ikd} - 2B + Ae^{-ikd}\right) \end{aligned} \right] \tag{4}$$

from which we can write

$$\left. \begin{aligned} A\left[\frac{2\tau}{m'd} - \omega^2\right] - B\frac{2\tau}{m'd}\cos kd &= 0 \\ -A\frac{2\tau}{m''d}\cos kd + B\left[\frac{2\tau}{m''d} - \omega^2\right] &= 0 \end{aligned} \right] \tag{5}$$

The solution to this set of coupled equations is obtained by setting the determinant of the coefficients equal to zero. We then obtain the secular equation

$$\left[\frac{2\tau}{m'd} - \omega^2\right]\left[\frac{2\tau}{m''d} - \omega^2\right] - \frac{1}{m'm''}\left[\frac{2\tau}{d}\cos kd\right]^2 = 0 \tag{6}$$

Solving for ω, we find

$$\omega^2 = \frac{\tau}{d}\left[\left(\frac{1}{m'} + \frac{1}{m''}\right) \pm \left[\left(\frac{1}{m'} + \frac{1}{m''}\right)^2 - \frac{4}{m'm''}\sin^2 kd\right]^{1/2}\right] \tag{7}$$

from which we find the two solutions

$$\omega_1^2 = \frac{\tau}{d}\left[\left(\frac{1}{m'}+\frac{1}{m''}\right)+\left[\left(\frac{1}{m'}+\frac{1}{m''}\right)^2-\frac{4}{m'm''}\sin^2 kd\right]^{1/2}\right]$$

$$\omega_2^2 = \frac{\tau}{d}\left[\left(\frac{1}{m'}+\frac{1}{m''}\right)-\left[\left(\frac{1}{m'}+\frac{1}{m''}\right)^2-\frac{4}{m'm''}\sin^2 kd\right]^{1/2}\right]$$

(8)

If $m' < m''$, and if we define

$$\omega_a \equiv \sqrt{\frac{2\tau}{m''d}}, \quad \omega_b \equiv \sqrt{\frac{2\tau}{m'd}}, \quad \omega_c = \sqrt{\omega_a^2+\omega_b^2} \tag{9}$$

Then the ω vs. k curve has the form shown below in which two branches appear, the lower branch being similar to that for $m' = m''$ (see Fig. 13-5).

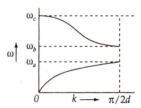

Using (9) we can write (6) as

$$\sin^2 kd = \frac{\omega^2}{\omega_a^2\omega_b^2}\left(\omega_a^2+\omega_b^2-\omega^2\right) \equiv W(\omega) \tag{10}$$

From this expression and the figure above we see that for $\omega > \omega_c$ and for $\omega_a < \omega < \omega_b$, the wave number k is complex. If we let $k = \kappa + i\beta$, we then obtain from (10)

$$\sin^2(\kappa+i\beta)d = \sin^2 \kappa d \cosh^2 \beta d - \cos^2 \kappa d \sinh^2 \beta d + 2i\sin \kappa d \cos \kappa d \sinh \beta d \cosh \beta d = W(\omega) \tag{11}$$

Equating the real and imaginary parts, we find

$$\left.\begin{array}{c}\sin \kappa d \cos \kappa d \sinh \beta d \cosh \beta d = 0\\[2mm]\sin^2 \kappa d \cosh^2 \beta d - \cos^2 \kappa d \sinh^2 \beta d = W(\omega)\end{array}\right] \tag{12}$$

We have the following possibilities that will satisfy the first of these equations:

a) $\sin \kappa d = 0$, which gives $\kappa = 0$. This condition also means that $\cos \kappa d = 1$; then β is determined from the second equation in (12):

$$-\sinh^2 \beta d = W(\omega) \tag{13}$$

Thus, $\omega > \omega_c$, and κ is purely imaginary in this region.

b) $\cos \kappa d = 0$, which gives $\kappa = \pi/2d$. Then, $\sin \kappa d = 1$, and $\cosh^2 \beta d = W(\omega)$. Thus, $\omega_a < \omega < \omega_b$, and κ is constant at the value $\pi/2d$ in this region.

c) $\sinh \beta d = 0$, **which gives** $\beta = 0$. Then, $\sin^2 \kappa d = W(\omega)$. Thus, $\omega < \omega_a$ or $\omega_b < \omega < \omega_c$, and κ is real in this region.

Altogether we have the situation illustrated in the diagram.

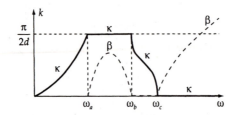

$$\gamma = \sqrt{1 - \tanh^2 \alpha} = \sqrt{1 - \frac{\sinh^2 \alpha}{\cosh^2 \alpha}} = \sqrt{\frac{\cosh^2\alpha - \sinh^2\alpha}{\cosh^2}} = \sqrt{\frac{1}{\cosh^2}} = \cosh^{\alpha}\alpha$$

The Special Theory
of Relativity

$$x' = (x_1 - vt)\gamma$$
$$= \cosh\alpha \left(x_1 - \frac{\sinh\alpha}{\cosh\alpha} ct \right)$$
$$= \cosh\alpha \, x_1 - \frac{ct}{\sinh\alpha}$$

$$v = \frac{\sinh\alpha}{\cosh\alpha} c$$

14-2. We introduce $\cosh \alpha \cong y$, $\sinh \alpha \cong y\, v/c$ and substitute these expressions into Eqs. (14.14); then

$$\left. \begin{aligned} x_1' &= x_1 \cosh \alpha - ct \sinh \alpha \\[2mm] t' &= t \cosh a - \frac{x_1}{c} \sinh \alpha \\[2mm] x_2' &= x_2; \qquad x_3' = x_3 \end{aligned} \right\} \tag{1}$$

Now, if we use $\cosh \alpha = \cos (i\alpha)$ and $i \sinh \alpha = \sin (i\alpha)$, we can rewrite (1) as

$$\left. \begin{aligned} x_1' &= x_1 \cos (i\alpha) + ict \sin (i\alpha) \\[2mm] ict' &= -x_1 \sin (i\alpha) + ict \cos (i\alpha) \end{aligned} \right\} \tag{2}$$

Comparing these equations with the relation between the rotated system and the original system in ordinary three-dimensional space,

$$\left. \begin{aligned} x_1' &= x_1 \cos \theta + x_2 \sin \theta \\[2mm] x_2' &= -x_1 \sin \theta + x_2 \cos \theta \\[2mm] x_3' &= x_3 \end{aligned} \right\} \tag{3}$$

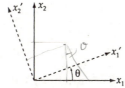

We can see that (2) corresponds to a rotation of the $x_1 - ict$ plane through the angle $i\alpha$.

14-8. The velocity of a point on the surface of the Earth at the equator is

$$v = \frac{2\pi R_e}{\tau} = \frac{2\pi \times \left(6.38 \times 10^8 \text{ cm}\right)}{8.64 \times 10^4 \text{ sec}}$$

$$= 4.65 \times 10^4 \text{ cm/sec} \qquad (1)$$

which gives

$$\beta = \frac{v}{c} = \frac{4.65 \times 10^4 \text{ cm/sec}}{3 \times 10^{10} \text{ cm/sec}} = 1.55 \times 10^{-6} \qquad (2)$$

According to Eq. (14.20), the relationship between the polar and equatorial time intervals is

$$\Delta t' = \frac{\Delta t}{\sqrt{1-\beta^2}} \cong \Delta t \left(1 + \frac{1}{2}\beta^2\right) \qquad (3)$$

so that the accumulated time difference is

$$\Delta = \Delta t' - \Delta t = \frac{1}{2}\beta^2 \Delta t \qquad (4)$$

Supplying the values, we find

$$\Delta = \frac{1}{2} \times \left(1.55 \times 10^{-6}\right) \times \left(3.156 \times 10^7 \text{ sec/yr}\right) \times \left(10^2 \text{ yr}\right) \qquad (5)$$

Thus,

$$\boxed{\Delta = 0.0038 \text{ sec}} \qquad (6)$$

14-11.

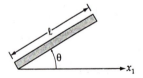

From example 14.1 we know that, to an observer in motion relative to an object, the dimensions of objects are contracted by a factor of $\sqrt{1-v^2/c^2}$ in the direction of motion. Thus, the x_1' component of the stick will be

$$\ell \cos \theta \sqrt{1-v^2/c^2}$$

while the perpendicular component will be unchanged:

$$\ell \sin \theta$$

So, to the observer in K', the length and orientation of the stick are

$$\ell' = \ell \left[\sin^2 \theta + \left(1-v^2/c^2\right) \cos^2 \theta \right]^{1/2}$$

$$\theta' = \tan^{-1}\left[\frac{\sin\theta}{\cos\theta\sqrt{1-v^2/c^2}}\right]$$

or

$$\ell' = \ell\left[\sin^2\theta + \frac{\cos^2\theta}{\gamma^2}\right]^{1/2}$$

$$\tan\theta' = \gamma\tan\theta$$

14-14.

K

source

K'

$v \leftarrow$

receiver

In K, the energy and momentum of each photon emitted are

$$E = h\nu_0 \quad \text{and} \quad p = \frac{h\nu_0}{c}$$

Using Eq. (14.92) to transform to K':

$$E' = h\nu = \gamma(E - vp_1); \quad \left(p_1 = -\frac{h\nu_0}{c}\right)$$

$$= \gamma\left(h\nu_0 + \frac{v}{c}h\nu_0\right)$$

So

$$\nu = \nu_0\,\gamma\left(1 + \frac{v}{c}\right)$$

$$= \nu_0\frac{1+\beta}{\sqrt{1-\beta^2}} = \nu_0\sqrt{\frac{1+\beta}{1-\beta}}$$

which agrees with Eq. (14.31).

14-19.

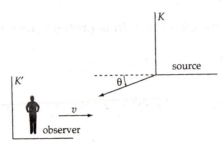

Proceeding as in the previous problem, we have

In K': $E' = h\nu$

$$p_1' = -\frac{h\nu}{c}\cos\theta = -\frac{h\nu}{c}\frac{\beta_r}{\sqrt{\beta_r^2 + \beta_t^2}}$$

In K: $E = \gamma(E' + vp_1') = h\nu_0$

So

$$h\nu_0 = \frac{1}{\sqrt{1 - \beta_r^2 - \beta_t^2}}\left[h\nu - \left[c\sqrt{\beta_r^2 + \beta_t^2}\right]\left[\frac{h\nu\,\beta_r}{c\sqrt{\beta_r^2 + \beta_t^2}}\right]\right]$$

or

$$\nu_0 = \frac{\nu(1 - \beta_r)}{\sqrt{1 - \beta_r^2 - \beta_t^2}}$$

$$\boxed{\frac{\nu}{\nu_0} = \frac{\lambda_0}{\lambda} = \frac{\sqrt{1 - \beta_r^2 - \beta_t^2}}{1 - \beta_r}}$$

For $\lambda > \lambda_0$, we have

$$(1 - \beta_r)^2 > 1 - \beta_r^2 - \beta_t^2$$

$$\beta_t^2 > 2\beta_r - 2\beta_r^2$$

$$\boxed{\beta_t^2 > 2\beta_r(1 - \beta_r)}$$

14-24.   The minimum energy will occur when the four particles are all at rest in the center of the mass system after the collision.

Conservation of energy gives (in the CM system)

$$2E_p = 4m_p c^2$$

or

$$E_{p,\text{CM}} = 2m_p c^2 = 2E_0$$

which implies $\gamma = 2$ or $\beta = \sqrt{3}/2$

To find the energy required in the lab system (one proton at rest initially), we transform back to the lab

$$E = \gamma\left(E' + vp_1'\right) \qquad (1)$$

The velocity of K'(CM) with respect to K(lab) is just the velocity of the proton in the K' system. So $u = v$.

Then

$$vp_1' = v\left(p_{CM}\right) = v\left(\gamma mu\right) = \gamma mv^2 = \gamma mc^2 \beta^2$$

Since $\gamma = 2$, $\beta = \sqrt{3}/2$,

$$vp_1' = \frac{3}{2}E_0$$

Substituting into (1)

$$E_{lab} = \gamma\left(2E_0 + \frac{3}{2}E_0\right) = 2\left[\frac{7}{2}E_0\right] = 7E_0$$

The minimum proton energy in the lab system is $7\,m_p c^2$, of which $6\,m_p c^2$ is kinetic energy.

14-28. $\quad T_{classical} = \dfrac{1}{2}mv^2$

$$T_{rel} = \left(\gamma - 1\right)mc^2 \geq T_{classical}$$

We desire

$$\frac{T_{rel} - T_{classical}}{T_{rel}} \leq 0.01$$

$$1 - \frac{\dfrac{1}{2}mv^2}{\left(\gamma - 1\right)mc^2} \leq 0.01$$

$$\frac{\dfrac{1}{2}v^2}{\left(\gamma - 1\right)c^2} \geq 0.99$$

$$\frac{\beta^2}{\gamma - 1} \geq 1.98$$

Putting $\gamma = \left(1 - \beta^2\right)^{-1/2}$ and solving gives

$$v \le 0.115\,c$$

> The classical kinetic energy will be within 1% of the correct value for $0 \le v \le 3.5 \times 10^7$ m/sec, independent of mass.

14-31.

Conservation of energy gives

$$E_\pi = 2E_p$$

where E_p = energy of each photon (Cons. of p_y implies that the photons have the same energy).

Thus

$$\gamma\,E_0 = 2E_p$$

$$E_p = \frac{\gamma\,E_0}{2} = \frac{135\ \text{MeV}}{2\sqrt{1 - 0.98^2}} = 339\ \text{MeV}$$

> The energy of each photon is 339 MeV.

Conservation of p_x gives

$$\gamma m v = 2p_p \cos\theta \quad \text{where } p_p = \text{momentum of each photon}$$

$$\cos\theta = \frac{\left(135\ \text{Mev}/c^2\right)(0.98\ c)}{2\sqrt{1 - 0.98^2}\,(339\ \text{MeV}/c)} = 0.98$$

> $\theta = \cos^{-1} 0.98 = 11.3°$

14-34.
$$\Delta s'^2 = -c^2 t'^2 + x_1'^2 + x_2'^2 + x_3'^2$$

Using the Lorentz transformation this becomes

$$\Delta s'^2 = \frac{-c^2 t^2 - \dfrac{v^2 x_1^2}{c^2} + 2x_1 vt}{1 - v^2/c^2} + \frac{x_1^2 + v^2 t^2 - 2x_1 vt}{1 - v^2/c^2} + x_2^2 + x_3^2$$

$$= \frac{\left[x_1^2 - \dfrac{v^2 x_1^2}{c^2} \right] - c^2 \left[t^2 - \dfrac{v^2}{c^2} t^2 \right]}{1 - v^2/c^2} + x_2^2 + x_3^2$$

$$= -c^2 t^2 + x_1^2 + x_2^2 + x_3^2$$

So

$$\boxed{\Delta s'^2 = \Delta s^2}$$

14-36. Since

$$F_\mu = \frac{d}{d\tau}\left[m \frac{dX\mu}{d\tau} \right] \text{ and } X_\mu = \left(x_1, x_2, x_3, ict \right)$$

we have

$$F_1 = \frac{d}{d\tau}\left[m \frac{dx_1}{d\tau} \right] = m \frac{d^2 x_1}{d\tau^2}$$

$$F_2 = m \frac{d^2 x_2}{d\tau^2} \qquad F_3 = m \frac{d^2 x_3}{d\tau^2}$$

$$F_4 = \frac{d}{d\tau}\left[m \frac{d(ict)}{d\tau} \right] = icm \frac{d^2 t}{d\tau^2}$$

Thus

$$F_1' = m\frac{d^2x_1}{d\tau^2} = m\frac{d^2}{d\tau^2}\left[\gamma(x_1 - vt)\right]$$

$$= \gamma m\frac{d^2x_1'}{d\tau^2} - \gamma mv\frac{d^2t}{d\tau^2} = \gamma\left(F_1 + i\beta F_4\right)$$

$$F_2' = m\frac{d^2x_2'}{d\tau^2} = m\frac{d^2x_2}{d\tau^2} = F_2 ; \quad F_3' = F_3$$

$$F_4' = icm\frac{d}{d\tau^2}\left[\gamma\left(t - \frac{vx_1}{c^2}\right)\right]$$

$$= \gamma\, icm\frac{d^2t}{d\tau^2} - \gamma\, i\beta m\frac{d^2x_1}{d\tau^2}$$

$$= \gamma\left(F_4 - i\beta F_1\right)$$

Thus the required transformation equations are shown.

14-41. We want to compute

$$\frac{T_1}{T_0} = \frac{E_1 - m_0c^2}{E_0 - m_0c^2} \tag{1}$$

where T and E represent the kinetic and total energy in the laboratory system, respectively, the subscripts 0 and 1 indicate the initial and final states, and m_0 is the rest mass of the incident particle.

The expression for E_0 in terms of γ_1 is

$$E_0 - m_0c^2\gamma_1 \tag{2}$$

E_1 can be related to E_1' (total energy of particle 1 in the center of momentum reference frame after the collision) through the Lorentz transformation [cf. Eq. (14.92)] (remembering that for the inverse transformation we switch the primed and unprimed variables and change the sign of v):

$$E_1 = \gamma_1'\left(E_1' + c\beta_1'p_1'\cos\theta\right) \tag{3}$$

where $p_1' = m_0c\beta_1'\gamma_1'$ and $E_1' = m_0c^2\gamma_1'$:

$$E_1 = m_0c^2\gamma_1'^2\left(1 + \beta_1'^2\cos\theta\right) \tag{4}$$

Then, from (1), (2), and (4),

$$\frac{T_1}{T_0} = \frac{\gamma_1'^2 + \gamma_1'^2\beta_1'^2\cos\theta - 1}{\gamma_1 - 1} \tag{5}$$

For the case of collision between two particles of equal mass, we have, from Eq. (14.127),

$$\gamma_1'^2 = \frac{1 + \gamma_1}{2} \tag{6}$$

and, consequently,

$$\gamma_1'^2 \beta_1'^2 = \gamma_1'^2 - 1 = \frac{\gamma_1 - 1}{2} \tag{7}$$

Thus, with the help of (6) and (7), (5) becomes

$$\frac{T_1}{T_0} = \frac{\gamma_1 - 1 + (\gamma_1 - 1)\cos\theta}{2(\gamma_1 - 1)}$$

$$= \frac{1 + \cos\theta}{2} \tag{8}$$

We must now relate the scattering angle θ in the center of momentum system to the angle ψ in the lab system.

Squaring Eq. (14.128), which is valid only for $m_1 = m_2$, we obtain an equation quadratic in $\cos\theta$. Solving for $\cos\theta$ in terms of $\tan^2\psi$, we obtain

$$\cos\theta = \frac{-\dfrac{\gamma_1 + 1}{2}\tan^2\psi \pm 1}{1 + \dfrac{\gamma_1 + 1}{2}\tan^2\psi} \tag{9}$$

One of the roots given in (9) corresponds to $\theta = \pi$, i.e., the incident particle reverses its path and is projected back along the incident direction. Substitution of the other root into (8) gives

$$\frac{T_1}{T_0} = \frac{1}{1 + \dfrac{\gamma_1 + 1}{2}\tan^2\psi} = \frac{2\cos^2\psi}{2\cos^2\psi + (\gamma_1 + 1)\sin^2\psi} \tag{10}$$

An elementary manipulation with the denominator of (10), namely,

$$2\cos^2\psi + (\gamma_1 + 1)\sin^2\psi = 2\cos^2\psi + \gamma_1(1 - \cos^2\psi) + \sin^2\psi$$

$$= \gamma_1 + \sin^2\psi + \cos^2\psi - \gamma_1\cos^2\psi + \cos^2\psi$$

$$= \gamma_1 + 1 - \gamma_1\cos^2\psi + \cos^2\psi$$

$$= (\gamma_1 + 1) - (\gamma_1 - 1)\cos^2\psi \tag{11}$$

provides us with the desired result:

$$\boxed{\frac{T_1}{T_0} = \frac{2\cos^2\psi}{(\gamma_1 + 1) - (\gamma_1 - 1)\cos^2\psi}} \tag{12}$$

Notice that the shape of the curve changes when $T_1 > m_0 c^2$, i.e., when $\gamma_1 > 2$.

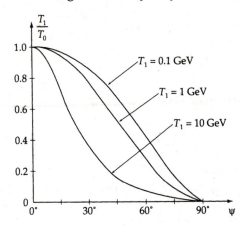